Jędrzej Śniatycki

Geometric Quantization and Quantum Mechanics

Springer-Verlag
New York Heidelberg Berlin

Jędrzej Śniatycki
The University of Calgary
2920 24th Avenue, N.W.
Calgary, Canada T2N 1N4

AMS Subject Classification: 81–02

Library of Congress Cataloging in Publication Data

Śniatycki, Jędrzej.
Geometric quantization and quantum mechanics.

(Applied mathematical sciences; v. 30)
Bibliography: p.
Includes index.
1. Quantum theory. 2. Geometry, Differential.
I. Title. II. Series.
QA1.A647 vol. 30 [QC174.12] 510'.8s [530.1'1]
79-26090

Printed in the United States of America.

9 8 7 6 5 4 3 2 1

ISBN 0-387-90469-7 Springer-Verlag New York Heidelberg Berlin
ISBN 3-540-90469-7 Springer-Verlag Berlin Heidelberg New York

To Rena

PREFACE

This book contains a revised and expanded version of the lecture notes of two seminar series given during the academic year 1976/77 at the Department of Mathematics and Statistics of the University of Calgary, and in the summer of 1978 at the Institute of Theoretical Physics of the Technical University Clausthal. The aim of the seminars was to present geometric quantization from the point of view of its applications to quantum mechanics, and to introduce the quantum dynamics of various physical systems as the result of the geometric quantization of the classical dynamics of these systems.

The group representation aspects of geometric quantization as well as proofs of the existence and the uniqueness of the introduced structures can be found in the expository papers of Blattner, Kostant, Sternberg and Wolf, and also in the references quoted in these papers. The books of Souriau (1970) and Simms and Woodhouse (1976) present the theory of geometric quantization and its relationship to quantum mechanics. The purpose of the present book is to complement the preceding ones by including new developments of the theory and emphasizing the computations leading to results in quantum mechanics.

I am greatly indebted to the participants of the seminars, in particular John Baxter, Eugene Couch, Jan Tarski, and Peter Zvengrowski, for encouragement and enlightening discussions, and to Bertram Kostant, John Rawnsley and David Simms for their interest in this work and their very helpful suggestions. Special thanks are due to Liisa Heikkila,

Pat Dalgetty, and Katherine MacDougall for careful and patient typing and retyping of the manuscript, and to Mark Gotay for editing the final draft of this work.

The work on this project was started during the 1976 Summer Research Institute of the Canadian Mathematical Congress (Victoria branch) and it was partially supported by the Natural Sciences and Engineering Research Council of Canada under grant No. A8091. During the summer of 1978 the author was a guest of the Institute of Theoretical Physics of the Technical University Clausthal and the hospitality of Prof. H.D. Doebner in Clausthal is gratefully acknowledged. The work on the final draft of the manuscript was facilitated by a Killam Resident Fellowship held by the author at the University of Calgary in Fall 1978.

Calgary, September, 1979 Jędrzej Śniatycki

CONTENTS

1. INTRODUCTION

1.1. Background

A classical system is described by the Poisson algebra of functions on the phase space of the system. Quantization associates to each classical system a Hilbert space $\mathscr{H}$ of quantum states and defines a map $\mathscr{Q}$ from a subset of the Poisson algebra to the space of symmetric operators on $\mathscr{H}$. The domain of $\mathscr{Q}$ consists of all "$\mathscr{Q}$-quantizable" functions. The definition of $\mathscr{Q}$ requires some additional structure on the phase space. The functions which generate one-parameter groups of canonical transformations preserving this additional structure are $\mathscr{Q}$-quantizable. They form a subalgebra of the Poisson algebra satisfying

$$[\mathscr{Q}f_1, \mathscr{Q}f_2] = i\hbar\mathscr{Q}[f_1,f_2],$$

where $[f_1,f_2]$ denotes the Poisson bracket of f_1 and f_2.

Two quantizations $\mathscr{Q}$ and $\mathscr{Q}'$ of the same classical system are *equivalent* if the domains of $\mathscr{Q}$ and $\mathscr{Q}'$ coincide and there exists a unitary operator $\mathscr{U}$ between the corresponding representation spaces such that, for each quantizable function f,

$$\mathscr{U}(\mathscr{Q}f) = (\mathscr{Q}'f)\mathscr{U} .$$

In physics, one is not interested in the whole Poisson algebra but rather in its subset consisting of functions with a definite physical interpretation, e.g., energy, momentum, and so on. Therefore, one may weaken the notion of equivalence of quantizations by requiring only that the physically interesting functions be contained in the intersection of the domains of $\mathscr{Q}$ and $\mathscr{Q}'$, and that the operator $\mathscr{U}$ intertwine the quantizations of these functions. This weaker notion of equivalence depends very much on our knowledge of the physical system under consideration and our judgement as to which functions are physically important.

There is a striking similarity between the canonical quantization of classical systems and the orbit method of construction of irreducible unitary representations of Lie groups. This similarity was recognized by Kostant, who wrote in the introduction of his 1970 paper entitled "Quantization and Unitary Representations":

> . . . We have found that when the notion of what the physicists mean by quantizing a function is suitably generalized and made rigorous, one may develop a theory which goes a long way towards constructing all the irreducible unitary representations of a connected Lie group. In the compact case it encompasses the Borel-Weil theorem. Generalizing Kirillov's result on nilpotent groups, L. Auslander and I have shown that it yields all the irreducible unitary representations of a solvable group of type I. (Also a

criterion for being of type I is simply expressed in terms of the theory.) For the semi-simple case, by results of Harish-Chandra and Schmid, it appears that enough representations are constructed this way to decompose the regular representation.

The geometric formulation of the canonical quantization scheme in physics was studied independently by Souriau. A comprehensive presentation of Souriau's theory of geometric quantization is contained in his book entitled "Structure des Systèmes Dynamiques" published in 1970. The works of Kostant and Souriau are the sources of the geometric quantization theory, also referred to as the "Kostant-Souriau theory."

The next fundamental development of the geometric quantization theory was due to Blattner, Kostant, and Sternberg [cf. Blattner (1973)]. It consists of the construction of a sesquilinear pairing between the representation spaces of the same classical system, usually referred to as a "Blattner-Kostant-Sternberg kernel." In some cases the pairing leads to the operator $\mathcal{U}$ intertwining the quantizations. As a result, one obtains a larger class of quantizable functions and the means of studying the equivalence of quantizations.

Geometric quantization is essentially a globalization of the canonical quantization scheme in which the additional structure needed for quantization is explicitly expressed in geometric terms. The theory, only about a decade old, is at a preliminary stage of its development. At present, it provides a unified framework for the quantization of classical systems which, when applied to most classical systems of physical interest, yields the expected quantum theories for these systems and removes some of the ambiguities left by other quantization schemes. It enables us to pose questions about the quantum theories corresponding to a given classical system and gives some partial answers. However, many issues remain unresolved. Among them are basic questions about the structure of the representation space, the search for appropriate conditions guaranteeing the convergence of the integrals involved in the Blattner-Kostant-Sternberg kernels and the unitarity of the intertwining operators defined by these kernels, etc. On a more specific level, there are cases when the geometric quantization of functions of physical interest poses such technical or theoretical difficulties that the corresponding quantum operators remain ambiguous. Some of these problems will be solved within the framework of the present theory. The others might require a modification of the theory; there are already indications that some modifications of the theory are inevitable.

The aim of this book is to present the theory of geometric quantization from the point of view of its applications to quantum mechanics, and to introduce the quantum dynamics of various physical systems as the result of the geometric quantization of the classical dynamics of these systems. It is assumed that the reader is familiar with classical and quantum mechanics and with the geometry of manifolds including the theory of connections. The proofs of the existence and the uniqueness of the structures introduced are omitted. On the other hand, all of the basic steps involved in computations are given, even though they may involve standard techniques.

A chapter by chapter description of the contents of the book follows.

1.2. Hamiltonian dynamics

A comprehensive exposition of classical mechanics containing references to the original papers is given by Whittaker (1961). The modern differential geometric approach adopted here follows Abraham and Marsden (1978).

The *phase space* of a dynamical system is a smooth manifold X endowed with a symplectic form ω defined by the Lagrange bracket. To each smooth function f on X, there is associated the *Hamiltonian vector field* ξ_f of f, defined by

$$\xi_f \rfloor \omega = -df,$$

as well as the one-parameter group $\phi_f^{\,t}$ of canonical transformations of (X,ω) generated by f which is obtained by integrating the vector field ξ_f. Define local coordinates $(q^1,\ldots,q^n,p_1,\ldots,p_n)$ on X, where $n = \frac{1}{2}\dim X$, such that

$$\omega = \sum_i dp_i \wedge dq^i.$$

In such a "canonical" chart, the integral curves of ξ_f satisfy the canonical equations of Hamilton with the Hamiltonian f.

The mapping $f \mapsto \xi_f$ pulls back the Lie algebra structure from the space of smooth vector fields on X to the space of smooth functions on X. The space of smooth functions on X with this induced Lie algebra structure is called the *Poisson algebra* of (X,ω).

The Hamiltonian formulation can be extended to relativistic dynamics. The Hamiltonian vector field of the mass-squared function yields the covariant form of the equations of motion. The interaction with an external electromagnetic field f is taken into account by adding the term ef to the symplectic form, where e is the charge of the particle. This approach to the relativistic dynamics of a charged particle is due to Souriau (1970). It has the advantage that it enables one to discuss the Hamiltonian dynamics of a relativistic charged particle without any reference to the electromagnetic potentials.

The evolution space formulation of Newtonian dynamics is due to Lichnerowicz (1943). For time-dependent dynamics, the evolution space formulation is more appropriate than the phase space formulation which requires a time-dependent Hamiltonian. The evolution space formulation of single particle dynamics is given following Śniatycki and Tulczyjew (1972); see also Souriau (1970).

1.3. Prequantization

In the first step of geometric quantization one associates, to each smooth function f on X, a linear operator $\mathcal{P}f$ such that $\mathcal{P}1$ is the identity operator and

$$[\mathcal{P}f, \mathcal{P}g] = i\hbar\,\mathcal{P}[f,g].$$

This is done by introducing a complex line bundle L over X with a connection ∇ and an invariant Hermitian form $\langle,\rangle$ such that

$$\text{curvature } \nabla = -h^{-1}\omega.$$

Such a line bundle exists if and only if $h^{-1}\omega$ defines an integral de Rham cohomology class. This condition, referred to as the *prequantization condition*, gives rise to the quantization of charge in Sec. 10.1 and spin in Sec. 11.2. The operators $\mathscr{P}f$ act on the space of sections of L as follows. The one-parameter group ϕ_f^t of canonical transformations generated by f has a unique lift to a one-parameter group of connection preserving transformations of L which defines the action of ϕ_f^t on the space of sections of L. The operator $\mathscr{P}f$ is then defined by

$$\mathscr{P}f[\lambda] = i\hbar \frac{d}{dt}(\phi_f{}^t\lambda)\Big|_{t=0}.$$

This definition also makes sense if f defines only a local one-parameter group of local canonical transformations.

For a function f on X such that the Hamiltonian vector field ξ_f is complete, the one-parameter group of linear transformations $\lambda \mapsto \phi_f^t$ preserves the scalar product given by

$$\langle\lambda_1|\lambda_2\rangle = \int_X \langle\lambda_1,\lambda_2\rangle\ \omega^n.$$

Hence, the operator $\mathscr{P}f$, defined originally on smooth sections of L, extends to a self-adjoint operator on the Hilbert space of square integrable sections of L. However, if we wanted to give a probabilistic interpretation to the scalar product by associating to $\langle\lambda,\lambda\rangle(x)$ the probability density of finding the "quantum" state described by λ in the classical state described by the point x in the phase space X, we would violate the uncertainty principle since square integrable sections of L can have arbitrarily small support. The space of all square integrable sections of L is too "big" to serve as the space of wave functions.

The prequantization of symplectic manifolds has been studied independently by Kostant (1970a) and Souriau (1970). Some physical implications of prequantization are discussed by Elhadad (1974), Kostant (1972), Rawnsley (1972; 1974), Renuard (1969), Simms (1972, 1973a,b), Souriau (1970), Streater (1967), and Śniatycki (1974). See also Sławianowski (1971, 1972) and Weinstein (1973).

The formulation of the theory of connections in complex line bundles given in Sec. 3.1 follows the general theory of connections given in Kobayashi and Nomizu (1963), modified by the identification of the complex line bundle without the zero section with the associated principal fibre bundle. The presentation of prequantization given in Sections 3.2 and 3.3 follows essentially the exposition of Kostant (1970a), where one may find the proofs of the theorems regarding the existence and the uniqueness of the prequantization structures.

1.4. Representation space

In order to reduce the prequantization representation one has to introduce a classical counterpart of a complete set of commuting observables. A first choice would be a set of n independent functions $f_1,\dots,f_n$ on X satisfying

$$[f_i,f_j] = 0 \quad \text{for } i,j = 1,2,\dots,n$$

such that their Hamiltonian vector fields are complete. The complex linear combinations of the Hamiltonian vector fields $\xi_{f_1},\dots,\xi_{f_n}$ give rise to a complex distribution F on X such that

$$[F,F] \subseteq F$$
$$\dim_{\mathbf{C}} F = \tfrac{1}{2} \dim X$$
$$\omega | F \times F = 0.$$

For many phase spaces of interest there does not exist such a set of functions. If one drops the assumption that the f_i be real and globally defined one is led to the notion of a *polarization* of (X,ω), that is, a complex distribution F on X satisfying the conditions given above. For technical reasons we assume that

$$D = F \cap \overline{F} \cap \mathscr{T}X$$

and

$$E = (F + \overline{F}) \cap \mathscr{T}X$$

are involutive distributions on X, and that the spaces X/D and X/E of the integral manifolds of D and E, respectively, are quotient manifolds of X with projections π_D and π_E. A polarization F satisfying these additional conditions is called *strongly admissible*.

Given a polarization F of (X,ω), one could take the space of sections of the prequantization line bundle L which are covariantly constant along F to form the representation space. However, if λ_1 and λ_2 are sections of L covariantly constant along F, their Hermitian product $\langle\lambda_1,\lambda_2\rangle$ is a function constant along D and its integral over X diverges unless the leaves of D are compact. Thus, we should integrate $\langle\lambda_1,\lambda_2\rangle$ over X/D, but we do not have a natural measure on X/D. In order to circumvent this difficulty, one introduces a bundle $\sqrt{\wedge^n F}$ sections of which can be paired to yield densities on X/D. The bundle $\sqrt{\wedge^n F}$ leads also to the correct modifications of the Bohr-Sommerfeld conditions and to unitary representations of groups of canonical transformations generated by certain dynamical variables.

The *representation space* $\mathcal{H}$ consists of those sections of $L \otimes \sqrt{\wedge^n F}$ which are covariantly constant along F and square integrable over X/D. The sections of $L \otimes \sqrt{\wedge^n F}$ covariantly constant along F have their supports in a subset S of X determined by the Bohr-Sommerfeld conditions; we will refer to S as the *Bohr-Sommerfeld variety* of the representation. If the Bohr-Sommerfeld conditions are nontrivial, one has to deal with distributional sections of $L \otimes \sqrt{\wedge^n F}$; this requires some further modifications of the scalar product.

If F is a *real* polarization, i.e.,

$$F = \overline{F} = D^c,$$

and the Hamiltonian vector fields in D are complete (*completeness condition)*, the structure of the representation space $\mathcal{H}$ is as follows: to each connected component S_α of the Bohr-Sommerfeld variety S, there corresponds a subspace $\mathcal{H}_\alpha$ of $\mathcal{H}$ consisting of sections with supports in S, and

$$\mathcal{H} = \oplus_\alpha \mathcal{H}_\alpha .$$

The representation space is trivial, $\mathcal{H} = 0$, if and only if the Bohr-Sommerfeld variety is empty.

The structure of the representation space for a non-real polarization has not been completely analyzed as yet. One has the following problem. Let M be an integral manifold of E contained in S. The restriction of $L \otimes \sqrt{\wedge^n F}$ to M induces a holomorphic line bundle $\tilde{L}_M$ over $\pi_D(M)$ in such a way that, to each section of $(L \otimes \sqrt{\wedge^n F})|M$ covariantly constant along $F|M$ there corresponds a unique holomorphic section of $\tilde{L}_M$. Let S' be the union of all integral manifolds M of E contained in S such that the holomorphic line bundle $\tilde{L}_M$ admits holomorphic sections which do not vanish identically. Then, every (discontinuous) section of $L \otimes \sqrt{\wedge^n F}$, provided its restrictions to integral manifolds of E are smooth and covariantly constant along F, vanishes outside S'. Under what conditions on the polarization F will the space of such sections admit a subspace of square integrable sections so that the resulting Hilbert space of wave functions is non-trivial if and only if $S' \neq \phi$? The conditions we are looking for are complex analogues of the completeness condition.

The notion of a real polarization was introduced independently by Kostant and Souriau. It corresponds to the Lagrangian foliation of a symplectic manifold studied by Weinstein (1971). Complex polarizations were used by Streater (1967), Shale and Stinespring (1967), and Kostant and Auslander (1971). The bundle $\sqrt{\wedge^n F}$ is closely related to the bundle of hal-forms normal to F introduced by Blattner (1973). The presentation of the structure of $\sqrt{\wedge^n F}$ given in Sec. 4.2 is patterned after Blattner's account of the structure of half-forms. The result that a strongly admissible polarization can be locally spanned by complex Hamiltonian vector fields follows from the complex Frobenius theorem of Nirenberg (1957). The completeness condition used here corresponds to the Pukansky condition in Auslander and Kostant (1971). The interpretation of the Bohr-Sommerfeld quantization conditions as conditions on the supports of the wave functions was given in Śniatycki (1975a). The structure the representation space defined by a complete strongly admissible real polarization, given in Sec. 4.5, follows from the results of Śniatycki and Toporowski (1977).

1.5. Blattner-Kostant-Sternberg kernels

Let F_1 and F_2 be two polarizations of (X,ω) and $\mathcal{H}_1$ and $\mathcal{H}_2$ the corresponding representation spaces. Under certain conditions there exists a geometrically defined sesquilinear map $\mathcal{K}: \mathcal{H}_1 \times \mathcal{H}_2 \to \mathbb{C}$ called the *Blattner-Kostant-Sternberg kernel*. The kernel $\mathcal{K}$ induces a linear map $\mathcal{U}: \mathcal{H}_2 \to \mathcal{H}_1$ such that

$$\mathcal{K}(\sigma_1,\sigma_2) = (\sigma_1 | \mathcal{U}\sigma_2),$$

where $(\cdot|\cdot)$ is the inner product on $\mathcal{H}_1$. If $\mathcal{U}$ is unitary, the representation spaces $\mathcal{H}_1$ and $\mathcal{H}_2$ are said to be *unitarily related.*

The construction of the Blattner-Kostant-Sternberg kernel requires the existence of a *metaplectic structure* for (X,ω), which is equivalent to the vanishing of the characteristic class in $H^2(X,\mathbb{Z}_2)$ of the bundle of symplectic frames of (X,ω). The set of all metaplectic structures for (X,ω), provided it is not empty, can be parametrized by $H^1(X,\mathbb{Z}_2)$, cf. Kostant (1973). A metaplectic structure on (X,ω) gives rise to the line bundle $\sqrt{\wedge^n F}$ for each polarization F satisfying the *positivity condition*

$$i\omega(w,\bar{w}) \geq 0$$

for all $w \in F$.

The kernel $\mathcal{K}(\sigma_1,\sigma_2)$ is defined by the integration of a density depending locally on the sections σ_1 and σ_2. The problem of determining appropriate conditions which guarantee the convergence of the integrals involved is as yet unsolved. The defining density is constructed here for pairs of positive polarizations F_1 and F_2 such that

$$D_{12} = D_1 \cap D_2$$

and

$$E_{12} = E_1 + E_2$$

are involutive distributions, and the spaces X/D_{12} and X/E_{12} of integral manifolds of D_{12} and E_{12}, respectively, are quotient manifolds of X.

The Blattner-Kostant-Sternberg kernel was introduced for transverse pairs of real polarizations in Blattner (1973). The definition for real non-transverse polarizations is due to Blattner (1975b). The construction of the kernel for regular pairs of positive polarizations was given also by Blattner (1977).

The definition of the positivity of a polarization used here is opposite to that of Blattner because the prequantization condition adopted in Chapter 3 requires that the curvature of the prequantization line bundle be $-\hbar^{-1}\omega$, rather than ω as in Blattner's papers. In the case of transverse pairs of positive polarizations, we follow the presentation of Blattner with obvious modifications due to some differences in terminology. We also allow for non-trivial Bohr-Sommerfeld conditions. For non-transverse pairs of positive polarizations our assumptions are more restrictive than the condition of regularity assumed by Blattner.

1.6. Quantization

Let $\mathscr{H}$ be the representation space corresponding to a polarization F. That is, $\mathscr{H}$ consists of sections of $L \otimes \sqrt{\wedge^n F}$ covariantly constant along F. Let f be a function on X such that its Hamiltonian vector field is complete, so that f generates a one-parameter group ϕ_f^t of canonical transformations of X. In quantizing f we have to distinguish two cases depending on whether or not ϕ_f^t preserves F.

If ϕ_f^t preserves F, its action can be lifted to the space of sections of $L \otimes \sqrt{\wedge^n F}$ covariantly constant along F. The *quantum operator* $\mathscr{Q}f$ *corresponding to* f is defined by

$$\mathscr{Q}f[\sigma] = i\hbar \frac{d}{dt}(\phi_f^t \sigma)\Big|_{t=0}$$

for each $\sigma \in \mathscr{H}$. Since the action of ϕ_f^t on $\mathscr{H}$ preserves the scalar product, $\mathscr{Q}f$ is a self-adjoint operator on $\mathscr{H}$. The definition of $\mathscr{Q}f[\sigma]$ given here also makes sense if the Hamiltonian vector field of f is incomplete, provided the support of σ projects to a compact set in X/D. Thus, the quantization of functions with incomplete Hamiltonian vector fields leads to densely defined symmetric operators on $\mathscr{H}$. The existence of a self-adjoint extension for such an operator has to be studied separately.

The functions f such that ϕ_f^t preserves F form a subalgebra of the Poisson algebra satisfying the commutation relations

$$[\mathscr{Q}f_1, \mathscr{Q}f_2] = i\hbar\, \mathscr{Q}[f_1, f_2].$$

There is an ideal of this subalgebra consisting of functions f such that the Hamiltonian vector fields ξ_f are contained in the polarization. For such a function f, the quantum operator $\mathcal{Q}f$ is the operator of multiplication by the function f:

$$\mathcal{Q}f[\sigma] = f\sigma$$

for each $\sigma \in \mathcal{A}$. The spectrum of $\mathcal{Q}f$ is contained in the image $f(S)$ of the Bohr-Sommerfeld variety S under the mapping $f: X \to \mathbb{R}$.

If $\phi_f{}^t$ does not preserve F, assume that it transforms F to a polarization $\mathcal{T}\phi_f{}^t(F)$ such that, for some $\varepsilon > 0$, the polarizations F and $\mathcal{T}\phi_f{}^t(F)$ satisfy the conditions for the existence of the Blattner-Kostant-Sternberg kernel $\mathcal{K}_t: \mathcal{A} \times \mathcal{A}_t \to \mathbb{C}$ whenever $0 < t < \varepsilon$. Here we have denoted by $\mathcal{A}_t$ the representation space corresponding to the polarization $\mathcal{T}\phi_f{}^t(F)$. For each $t \in (0,\varepsilon)$ there exists a linear map $\mathcal{U}_t: \mathcal{A}_t \to \mathcal{A}$ satisfying

$$\mathcal{K}(\sigma,\sigma_t) = (\sigma|\mathcal{U}_t\sigma_t)$$

for all $\sigma \in \mathcal{A}$ and $\sigma_t \in \mathcal{A}_t$. The one-parameter group $\phi_f{}^t$ of canonical transformations of X generated by f acts on sections of $L \otimes \sqrt{\wedge^n F}$ mapping them to sections of $L \otimes \sqrt{\wedge^n \mathcal{T}\phi_f{}^t(F)}$. The quantum operator $\mathcal{Q}f$ corresponding to f is then defined by

$$\mathcal{Q}f[\sigma] = i\hbar \frac{d}{dt}(\mathcal{U}_t \circ \phi_f{}^t\sigma)\Big|_{t=0}.$$

If, for each $t \in (0,\varepsilon)$, the operator $\mathcal{U}_t$ is unitary, then $\mathcal{Q}f$ is self-adjoint. In practice it may be difficult to verify the unitarity of $\mathcal{U}_t$, but one may still use the expression for $\mathcal{Q}f[\sigma]$ given here to evaluate $\mathcal{Q}f$ on a dense domain and investigate the existence of a self-adjoint extension afterwards.

The representation space $\mathscr{H}$ is the direct sum of the subspaces $\mathscr{H}_\alpha$ consisting of sections with supports in the connected components S_α of the Bohr-Sommerfeld variety S. If f is a function preserving F then, for each connected component S_α of S, $\phi_f^{\,t}(S_\alpha) = S_\alpha$ and the operator $\mathscr{Q}f$ maps $\mathscr{H}_\alpha$ to itself. If f does not preserve F but is quantizable by means of the Blattner-Kostant-Sternberg kernels, $\mathscr{Q}f$ also maps $\mathscr{H}_\alpha$ to itself; this can be seen as follows. Let σ be an element of $\mathscr{H}_\alpha$ with support projecting onto a compact set in X/D. Then, for each component S_β of S different than S_α, there exists $\delta > 0$ such that

$$\phi_f^{\,t}(\text{support } \sigma) \cap S_\beta = \phi$$

for all $0 < t < \delta$. Hence,

$$(\mathscr{U}_t \circ \phi_f^{\,t}\sigma \,|\, \sigma') = 0$$

for all $\sigma' \in \mathscr{H}_\beta$ and $t \in (0,\delta)$. Differentiating this equation with respect to t and setting $t = 0$ we get

$$(\mathscr{Q}f[\sigma] \,|\, \sigma') = 0$$

for each $\sigma' \in \mathscr{H}_\beta$. Hence

$$\mathscr{Q}f[\mathscr{H}_\alpha] \subseteq \mathscr{H}_\alpha$$

for each connected component S_α of S.

Thus, if all the observables in the quantum theory described in terms of the representation space $\mathscr{H}$ are obtained by the quantization of functions on X in the manner described here, we are led to *superselection rules*. The representation space decomposes into a direct sum of orthogonal subspaces corresponding to different components of the

Bohr-Sommerfeld variety, and all observables commute with the operators of projection onto these subspaces.

The presentation of geometric quantization given here essentially follows that of Blattner (1973). Superselection rules in geometric quantization are discussed in Śniatycki (1978b).

The assumptions on the polarizations F and $\mathcal{T}\phi_f{}^t(F)$ made here are somewhat too restrictive from the point of view of applications. In certain cases of physical interest the intersection $F \cap \mathcal{T}\phi_f{}^t(\overline{F})$ is not a distribution, and one will have to extend the construction of Blattner-Kostant-Sternberg kernels to cover this situation.

If a function cannot be quantized in the representation given by the polarization F, one can quantize with the help of an additional polarization. Let f be a function quantizable in the representation given by a polarization F'. We denote by $\mathcal{A}'$ the corresponding representation space and by $\mathcal{Q}'f$ the operator on $\mathcal{A}'$ corresponding to f. Suppose the polarizations F and F' allow for the existence of the Blattner-Kostant-Sternberg kernel $\mathcal{K}: \mathcal{A}' \times \mathcal{A} \to \mathbf{C}$, which in turn yields a linear isomorphism $\mathcal{U}: \mathcal{A}' \to \mathcal{A}$. Then $\mathcal{U}(\mathcal{Q}'f)\mathcal{U}^{-1}$ is a linear operator on $\mathcal{A}$ which could be taken for the quantum operator corresponding to the classical variable f. It should be noted that the operator $\mathcal{U}(\mathcal{Q}'f)\mathcal{U}^{-1}$ will depend in general on the choice of the auxiliary polarization F', and it need not be self-adjoint if $\mathcal{U}$ is not unitary. This technique has been used by Rawnsley (1977b) in the quantization of the geodesic flow on spheres.

1.7. Schrödinger representation

The *Schrödinger representation* is given by the polarization F spanned by the Hamiltonian vector fields of the position variables. For a single particle, there exists a section $\lambda_0 \otimes \nu_{\tilde{\underline{\xi}}}$ of $L \otimes \sqrt{\wedge^3 F}$ covariantly constant along F such that every element $\sigma \in \mathscr{H}$ is of the form

$$\sigma = \psi(\underline{q})\lambda_0 \otimes \nu_{\tilde{\underline{\xi}}},$$

where $\psi(\underline{q})$ is a square integrable function of the position variables $\underline{q}$, and the scalar product on $\mathscr{H}$ is given by

$$(\psi(\underline{q})\lambda_0 \otimes \nu_{\tilde{\underline{\xi}}} | \psi'(\underline{q})\lambda_0 \otimes \nu_{\tilde{\underline{\xi}}}) = \int_{R^3} \psi(\underline{q})\overline{\psi}'(\underline{q})d^3q.$$

This establishes in isomorphism between the representation space $\mathscr{H}$ and the space of square integrable complex-valued functions on R^3.

The physically interesting dynamical variables preserving the polarization F are the components of the position vector $\underline{q}$, the momentum vector $\underline{p}$, and the angular momentum vector $\underline{J}$. We obtain

$$\mathscr{Q}q^i[\psi(\underline{q})\lambda_0 \otimes \nu_{\tilde{\underline{\xi}}}] = q^i\psi(\underline{q})\lambda_0 \otimes \nu_{\tilde{\underline{\xi}}}$$

$$\mathscr{Q}p_i[\psi(\underline{q})\lambda_0 \otimes \nu_{\tilde{\underline{\xi}}}] = -i\hbar \frac{\partial}{\partial q^i}\psi(\underline{q})\lambda_0 \otimes \nu_{\tilde{\underline{\xi}}}$$

$$\mathscr{Q}J_i[\psi(\underline{q})\lambda_0 \otimes \nu_{\tilde{\underline{\xi}}}] = -i\hbar \sum_{jk} \varepsilon_{ijk}q^j \frac{\partial}{\partial q^k}\psi(\underline{q})\lambda_0 \otimes \nu_{\tilde{\underline{\xi}}}$$

in agreement with the Schrödinger theory. The energy

$$H = \underline{p}^2/2m + V(\underline{q})$$

generates a one-parameter group $\phi_H{}^t$ of canonical transformations which does not preserve the polarization. Hence, the quantization of H requires the existence of the Blattner-

Kostant-Sternberg kernels $\mathscr{K}_t : \mathscr{A} \times \mathscr{A}_t \to \mathbb{C}$, where $\mathscr{A}_t$ is the representation space associated to the polarization $\mathscr{T}\phi_H{}^t(F)$. Denoting by $\mathscr{U}_t : \mathscr{A}_t \to \mathscr{A}$ the linear map induced by $\mathscr{K}_t$, we obtain

$$\mathscr{U}_t \circ \phi_H{}^t[\psi(\underline{q})\lambda_o \otimes \nu_{\tilde{\underline{\xi}}}] = \left\{ \int_{R^3} d^3 q_o \psi(\underline{q}_o) K(\underline{q}_o; t, \underline{q}) \right\} \lambda_o \otimes \nu_{\tilde{\underline{\xi}}},$$

where the integral kernel $K(\underline{q}_o; t, \underline{q})$ is given by the Van Vleck approximation

$$K(\underline{q}_o; t, \underline{q}) = (-i\hbar)^{-3/2} \left\{ \det\left[\frac{\partial^2 S(0, \underline{q}_o; t, \underline{q})}{\partial q^j \, \partial q_o{}^k}\right]\right\}^{\frac{1}{2}} \exp[i\hbar^{-1} S(0, \underline{q}_o; t, \underline{q})].$$

Here $S(0, \underline{q}_o; t, \underline{q})$ is the classical action between the points $\underline{q}_o$ and $\underline{q}$ during the time interval $[0,t]$. Differentiating $\mathscr{U}_t \circ \phi_H{}^t$ with respect to t and setting $t = 0$, we obtain the Schrödinger expression for the energy operator

$$\mathscr{Q}H[\psi(\underline{q})\lambda_o \otimes \nu_{\tilde{\underline{\xi}}}] = [(-\hbar^2/2m)\Delta + V(\underline{q})]\psi(\underline{q})\lambda_o \otimes \nu_{\tilde{\underline{\xi}}}.$$

On the other hand, iteration of the expression for $\mathscr{U}_t \circ \phi_H{}^t[\psi(\underline{q})\lambda_o \otimes \nu_{\tilde{\underline{\xi}}}]$ leads to the Feynman path integral.

These results can be extended to a system of m particles with constraints. The phase space of the system is the contangent bundle space $\mathscr{T}^*Y$ of the configuration manifold $Y \subseteq R^{3m}$ defined by the constraints. The *Lagrange bracket* is given by

$$\omega = d\theta_Y,$$

where θ_Y is the *canonical 1-form* on $\mathscr{T}^*Y$ defined as follows. For each $x \in \mathscr{T}^*Y$ and each vector u in the tangent space $\mathscr{T}_x(\mathscr{T}^*Y)$, set

$$\theta_Y(u) = x(\mathscr{T}\pi(u)),$$

where $\pi : \mathscr{T}^*Y \to Y$ is the cotangant bundle projection and

$\mathcal{T}\pi: \mathcal{T}(\mathcal{T}^*Y) \to \mathcal{T}Y$ is the derived map. The polarization F is spanned by the vectors tangent to the fibres of $\pi: \mathcal{T}^*Y \to Y$.

The dynamical variables of physical interest are the energy H, the canonical coordinates $q^1, \ldots, q^n$, and the corresponding canonical momenta $p_1, \ldots, p_n$. However, the canonical coordinates and momenta are defined only locally. Since the geometric quantization scheme applies only to globally defined functions, we have to redefine our notions of the canonical positions and momenta. The characteristic property of the canonical coordinates $q^1, \ldots, q^n$ is that they are constant along the fibres of $\mathcal{T}^*Y$. Thus, we say that a function q on $\mathcal{T}^*Y$ is of "position type" if

$$q = \check{q} \circ \pi$$

for some function $\check{q}$ on Y. If ζ is a smooth vector field on Y, we define a function p_ζ on $\mathcal{T}^*Y$ by

$$p_\zeta(x) = x(\zeta(\pi(x)))$$

for each $x \in \mathcal{T}^*Y$. In particular, when $\zeta = \partial/\partial\check{q}^i$, p_ζ coincides with the canonical momentum p_i. For this reason we refer to p_ζ as "the canonical momentum associated to the vector field ζ." The kinetic energy K of the system defines a Riemannian metric g on Y such that

$$K(x) = \frac{1}{2} g(x,x)$$

for each $x \in \mathcal{T}^*Y$. If the potential energy is given by a function $V: Y \to \mathbf{R}$, the total energy is

$$H = K + V \circ \pi.$$

There exists a section $\lambda_o \otimes \nu_g$ of $L \otimes \sqrt{\wedge^n F}$ covariantly constant along F such that every section $\sigma \in \mathcal{A}$

is of the form

$$\sigma = \Psi \otimes \lambda_o \otimes \nu_g,$$

where Ψ is a complex-valued function on Y square integrable with respect to the density $|\det g|^{\frac{1}{2}}$ on Y defined by the Riemannian metric g. The scalar product on $\mathscr{A}$ is given by

$$(\Psi_1 \otimes \lambda_o \otimes \nu_g | \Psi_2 \otimes \lambda_o \otimes \nu_g) = \int_Y \Psi_1 \overline{\Psi}_2 |\det g|^{\frac{1}{2}}.$$

Hence, the mapping associating to each $\sigma = \Psi \otimes \lambda_o \otimes \nu_g$ the function Ψ on Y is a unitary isomorphism of the representation space $\mathscr{A}$ with the space of complex-valued functions on Y square integrable with respect to the density $|\det g|^{\frac{1}{2}}$.

The quantization of a position type function $q = \check{q} \circ \pi$ yields

$$\mathscr{Q}q[\Psi \otimes \lambda_o \otimes \nu_g] = \check{q}\Psi \otimes \lambda_o \otimes \nu_g.$$

For each smooth vector field ζ on Y, we obtain

$$\mathscr{Q}p_\zeta[\Psi \otimes \lambda_o \otimes \nu_g] = -i\hbar[\zeta\Psi + \tfrac{1}{2}(\mathrm{Div}\ \zeta)\Psi] \otimes \lambda_o \otimes \nu_g,$$

where $\mathrm{Div}\ \zeta$ denotes the covariant divergence of the vector field ζ. We see that the operator of the momentum associated to a vector field ζ corresponds to $-i\hbar$ times the operator of differentiation in the direction ζ if and only if

$$\mathrm{Div}\ \zeta = 0.$$

This condition is equivalent to the condition that the local one-parameter group generated by ζ preserve the metric density $|\det g|^{\frac{1}{2}}$. The quantization of energy proceeds, as before, via the Blattner-Kostant-Sternberg kernels. It yields

$$\mathscr{Q}H[\Psi \otimes \lambda_o \otimes \nu_g] = [-\frac{\hbar^2}{2}(\Delta\Psi - \frac{1}{6}R\Psi) + V\Psi] \otimes \lambda_o \otimes \nu_g$$

where Δ is the Laplace-Beltrami operator defined by the metric g and R is the scalar curvature of the metric connection on Y. It should be noted that the validity of this result depends on the convergence of the integrals defining the Blattner-Kostant-Sternberg kernels, and thus places some restrictions on the geometry of the configuration space. Iteration of the one-parameter family of transformations of $\mathscr{H}$ given by the Blattner-Kostant-Sternberg kernels leads to the Feynman path integral as in the case of a single particle.

The geometric quantization of the energy of a free particle in the Schrödinger representation is due to Blattner (1973). It led him to introduce the bundle of half-forms normal to the polarization in order to obtain the correct cancellation of factors in the method of stationary phase. The quantization of the kinetic energy of a system with a Riemannian configuration space was studied by Blattner (1973) and by Simms and Woodhouse (1976). The result obtained here differs from both their results by the correction term $-R/6$ to the Laplace-Beltrami operator which is due to the half-form nature of the wave functions. The appearance of a correction term proportional to the scalar curvature was first derived by DeWitt (1957). The computations in terms of normal coordinate systems used here follow those of Ben-Abraham and Lonke (1973), but the physical interpretation of the result is different. The properties of normal coordinates essential for our computations can be found in Schouten (1954), pp. 155-165.

The relationship between geometric quantization and the Feynman path integral formulation of quantum mechanics was studied by Blackman (1976), Gawędzki (1977) and Simms (1978b). A comprehensive discussion of quantum mechanics on a manifold from the point of view of the Feynmann path integral formalism can be found in Dowker (1975); see also the references quoted there.

1.8. Other representations

The choice of representation is determined by the choice of polarization. In the Schrödinger representation the polarization is spanned by the Hamiltonian vector fields of the positions. The *momentum representation* corresponds to the polarization spanned by the Hamiltonian vector fields of the momentum variables. The Blattner-Kostant-Sternberg kernel between these representations corresponds to the Fourier transform.

The *Bargmann-Fock representation* is given by the polarization F_B spanned by the Hamiltonian vector fields of the complex coordinates

$$z_k = 2^{-\frac{1}{2}}(p_k + iq^k).$$

There exists a section $\lambda_1 \otimes \nu_{\tilde{\xi}_{\underline{z}}}$ of $L \otimes \sqrt{\wedge^3 F_B}$ covariantly constant along F_B such that each element σ of the representation space $\mathcal{H}_B$ can be written

$$\sigma = \Psi(\underline{z})\lambda_1 \otimes \nu_{\tilde{\xi}_{\underline{z}}},$$

where Ψ is a holomorphic function of the complex coordinates $\underline{z} = (z_1, z_2, z_3)$. The scalar product on $\mathcal{H}_B$ is given by

$$(\Psi\lambda_1 \otimes \nu_{\tilde{\xi}_{\underline{z}}} | \Psi'\lambda_2 \otimes \nu_{\tilde{\xi}_{\underline{z}}}) = \int_{R^6} \Psi(\underline{z})\overline{\Psi}'(\underline{z})\exp(-|\underline{z}|^2/\hbar)d^3p\,d^3q.$$

Hence, the representation space $\mathcal{H}_B$ is isomorphic to the space of holomorphic functions on $\mathbb{C}^3$ with the scalar product given by the right hand side of the above equation.

As in the case of the Schrödinger representation, the components of the position, the momentum, and the angular momentum vectors preserve the polarization. Their quantization yields

$$\mathcal{Q}_B p_k [\Psi\lambda_1 \otimes \nu_{\tilde{\xi}_{\underline{z}}}] = 2^{-\frac{1}{2}}[(z_k + \hbar \frac{\partial}{\partial z_k})\Psi]\lambda_1 \otimes \nu_{\tilde{\xi}_{\underline{z}}}$$

$$\mathcal{Q}_B q^k [\Psi\lambda_1 \otimes \nu_{\tilde{\xi}_{\underline{z}}}] = -i2^{-\frac{1}{2}}[(z_k - \hbar \frac{\partial}{\partial z_k})\Psi]\lambda_1 \otimes \nu_{\tilde{\xi}_{\underline{z}}}$$

$$\mathcal{Q}_B J_k [\Psi\lambda_1 \otimes \nu_{\tilde{\xi}_{\underline{z}}}] = i\hbar \sum_{jm} \varepsilon_{kjm}\, z_j \frac{\partial \Psi}{\partial z_m} \lambda_1 \otimes \nu_{\tilde{\xi}_{\underline{z}}}$$

where we have used the symbol $\mathcal{Q}_B$ to denote the quantization in the Bargmann-Fock representation. The energy

$$H = \frac{1}{2} \sum_k [(q^k)^2 + (p_k)^2]$$

of a harmonic oscillator with unit mass and spring constant also preserves the polarization F_B. Its quantization yields

$$\mathcal{Q}_B H [\Psi\lambda_1 \otimes \nu_{\tilde{\xi}_{\underline{z}}}] = \hbar(\sum_k z_k \frac{\partial \Psi}{\partial z_k} + \frac{3}{2}\Psi)\lambda_1 \otimes \nu_{\tilde{\xi}_{\underline{z}}} .$$

The eigenvalues E_n of $\mathcal{Q}_B H$ are easily found to be of the form

$$E_n = (n + \tfrac{3}{2})\hbar \quad \text{for} \quad n = 0,1,2,\ldots .$$

It should be noted that the ground state energy $3\hbar/2$ is due to the transformation properties of the bundle $\sqrt{\wedge^3 F_B}$ under the one-parameter group of canonical transformations generated by H.

The polarization F_B is transverse to the polarization F which gives rise to the Schrödinger representation. The Blattner-Kostant-Sternberg kernel defined by the polarizations

F and F_B induces a unitary isomorphism $\mathcal{U}_B$ between $\mathcal{H}_B$ and the Schrödinger representation space $\mathcal{H}$ which intertwines the quantization of positions, momenta, angular momenta, and harmonic oscillator energy.

The representation of the canonical commutation relations on the space of holomorphic functions on the phase space was introduced by Fock (1928). The mathematics of this theory and its equivalence to the Schrödinger representation were analyzed by Bargmann (1961). An approach to the representation on the space of holomorphic functions from the standpoint of geometric quantization can be found in Guillemin and Sternberg (1977) Ch. V, §7.

The energy H of a one-dimensional harmonic oscillator can be quantized in the representation given by the polarization F_H spanned by the Hamiltonian vector field ξ_H of H. Since ξ_H vanishes at $p = q = 0$, the distribution spanned by ξ_H has a singularity at the origin. At present we have no theory of quantization for polarizations with singularities, so in order to avoid the singularity we remove the origin from the phase space. The representation space $\mathcal{H}_H$ consists of sections of $L \otimes \sqrt{\wedge^1 F_H}$ covariantly constant along F_H. The supports of such sections are contained in the Bohr-Sommerfeld variety S, which is the union of those concentric circles S_n satisfying the modified Bohr-Sommerfeld conditions

$$\oint_{S_n} p\,dq = (n + \tfrac{1}{2})h.$$

The correction term $\tfrac{1}{2}h$ is due to the choice of the trivial metaplectic structure for X and the fact that it induces a non-trivial metalinear frame bundle for F. For each $n = 0,1,2,\ldots$, the subspace $\mathscr{A}_n$ of $\mathscr{A}$ consisting of sections with supports in S_n is one-dimensional and is the eigenspace of the quantized Hamiltonian corresponding to the eigenvalue $(n + \frac{1}{2})\hbar$.

The quantization of the position and the momentum variables, which is so straightforward in the Schrödinger and the Bargmann-Fock representations, cannot be obtained at present in the energy representation for the following reason. Let $\phi_p{}^t$ denote the local one-parameter group of local canonical transformations generated by the canonical momentum p; it is only a local group since we have removed the origin from the phase space. The transformations $\phi_p{}^t$ correspond to translations along the q-axis. Since the integral curves of ξ_H are the circles $p^2 + q^2 = \text{const.}$, the distributions F_H and $\mathscr{T}\phi_p{}^t(F_H)$ do not intersect along a distribution. Hence, the polarizations F_H and $\mathscr{T}\phi_p{}^t(F_H)$ do not satisfy the conditions necessary for the construction of the Blattner-Kostant-Sternberg kernels. It may be that this is only a technical difficulty and in the future one will be able to generalize the construction of the kernels to the case of polarizations with singular intersections. However, if this generalization is local, i.e., the kernel $\mathscr{K}(\sigma_1,\sigma_2)$ is the integral of a concomitant $<\sigma_1,\sigma_2>$ depending locally on the sections σ_1 and σ_2, then the resulting operator $\mathscr{Q}_H p$ would commute with $\mathscr{Q}_H H$ according to the argument leading to the superselection rules. Thus, either the generalization of the Blattner-Kostant-Sternberg kernel used for quantizating the momentum in this energy representation is non-local or the operators $\mathscr{Q}_H p$ and $\mathscr{Q}_H H$ commute, in which case the energy representation is inequivalent to the Schrödinger representation. The same argument is valid for any dynamical variable f.

The Blattner-Kostant-Sternberg construction leads to a linear isomorphism $\mathcal{U}$ between $\mathcal{A}_H$ and the Bargmann-Fock representation space for one degree of freedom. The operator $\mathcal{U}$ intertwines the energy quantizations, but it fails to be unitary. One could make $\mathcal{U}$ unitary by redefining the scalar product in $\mathcal{A}_H$.

The geometric quantization of the harmonic oscillator in the energy representation was analyzed by Simms (1975a). The argument that one should use the trivial metaplectic structure induced by the unique metaplectic structure on R^2 is due to Blattner.

1.9. Time-dependent Schrödinger equation

The canonical formulation of dynamics is the usual starting point for the quantization of a classical system. However, in the case of non-relativistic dynamics the phase space of the canonical formulation is defined in terms of an inertial frame. Hence, one has to investigate the Galilei covariance of the theory. The evolution space formulation of non-relativistic dynamics allows for an intrinsic formulation of quantum dynamics independent of any inertial frame.

The *Galilean space-time* Y is a 4-dimensional affine space. The notion of simultaneity defines an affine mapping $\tau: Y \to T$, where T is a 1-dimensional oriented affine space representing absolute time. Each fibre of τ is a 3-dimensional Euclidean space. The translations in Y induce isometries of the fibres of τ. An *inertial frame* corresponds to an affine trivialization of τ which yields isometries of the fibres.

For each $t \in T$, the fibre Y_t of τ over t represents the physical space at time t. The phase space at time t is the cotangent bundle space $\mathcal{T}^*Y_t$ of Y_t, and the Lagrange bracket is given by the canonical 2-form of $\mathcal{T}^*Y_t$ denoted ω_t. The *evolution space* Z is the union of all phase spaces at a given time t as t varies over T,

$$Z = \bigcup_{t \in T} \mathcal{T}^*Y_t .$$

The Newtonian dynamics of a system is determined by an extension of the collection $\{\omega_t | t \in T\}$ of forms on the fibres of $Z \to T$ to a 2-form Ω on Z. Let N denote the characteristic distribution of Ω,

$$N = \{v \in \mathcal{T}Z \mid v \rfloor \Omega = 0\}.$$

The distribution N is one-dimensional, hence involutive. For each $z \in Z$, the evolution of the classical state z is given by the integral manifold of N through z. This description of dynamics can be put into Hamiltonian form if and only if Ω is closed. In the following we therefore assume that the dynamics under consideration is given by a closed form Ω.

We consider a complex line bundle L over Z with a connection ∇ and an invariant Hermitian form such that

$$\text{curvature } \nabla = -\hbar^{-1}\Omega.$$

For each $t \in T$, the restriction of L to $\mathcal{T}^*Y_t$ is a prequantization line bundle for $(\mathcal{T}^*Y_t, \omega_t)$. We denote by $\mathcal{H}_t$ the Schrödinger representation space for $(\mathcal{T}^*Y_t, \omega_t)$. The quantum evolution space is the union of the representation spaces corresponding to all $t \in T$,

$$\mathscr{S} = \bigcup_{t\in T} \mathscr{A}_t .$$

It is a bundle of Hilbert spaces over T. The quantum dynamics of the system is given by a trivialization of this bundle.

In order to describe the quantum dynamics induced by the classical dynamics defined by Ω, consider a vector field ζ in N which projects onto a constant vector field on T. The choice of ζ is equivalent to fixing the time scale. The vector field ζ_* is complete, and we denote by ϕ^s the one-parameter group of diffeomorphisms of Z generated by ζ. The group ϕ^s preserves Ω and it induces a one-parameter group of translations of T denoted by $t \mapsto t+s$. For each $t \in T$ and $s \in \mathbb{R}$, the transformation ϕ^s induces a unitary isomorphism $\phi_t^s\colon \mathscr{A}_{t-s} \to \mathscr{A}_t^s$, where $\mathscr{A}_t^s$ is the representation space for $(\mathscr{T}^*Y_t, \omega_t)$ corresponding to the image of the Schrödinger polarization under the transformation ϕ^s. Pairing the representation spaces $\mathscr{A}_t$ and $\mathscr{A}_t^s$ by means of the Blattner-Kostant-Sternberg kernel, one obtains a linear map $\mathscr{U}_{t,s}\colon \mathscr{A}_t^s \to \mathscr{A}_t$. Let $\Phi_s\colon \mathscr{S} \to \mathscr{S}$ be the bundle map such that, for each $t \in T$,

$$\Phi_s | \mathscr{A}_{t-s} = \mathscr{U}_{t,s} \circ \Phi_t^s .$$

The quantum dynamics is given as follows: a section σ of $\mathscr{S}$ describes a dynamically admissible history of the quantum system under consideration if and only if

$$\frac{d}{ds}(\Phi_s \sigma)\Big|_{s=0} = 0.$$

Introducing an inertial frame, one can show that this condition is equivalent to the time-dependent Schrödinger equation provided the conditions for the existence of the Blattner-Kostant-Sternberg kernels are satisfied.

The geometric quantization of time-dependent Hamiltonian systems was first studied by Simms (1978a). The intrinsic formulation given here follows that of Śniatycki (1978c).

1.10. Relativistic dynamics in an electromagnetic field

The relativistic dynamics of a particle with charge e in an external gravitational field represented by a metric g and an external electromagnetic field f is described by the phase space $(\mathcal{T}^*Y, \omega_e)$, where Y is the space-time manifold,

$$\omega_e = d\theta_Y + e\pi^*f,$$

and π^*f denotes the pull-back of the 2-form f on Y to $\mathcal{T}^*Y$. Let L_e be a prequantization line bundle corresponding to the symplectic form ω_e. The prequantization condition

$$\text{curvature } \nabla = -h^{-1}\omega_e$$

can be satisfied if and only if the de Rham cohomology class in $H^2(Y,\mathbf{R})$ defined by $h^{-1}ef$ is integral. This implies that, for each compact oriented 2-surface Σ in Y with empty boundary, the number

$$\int_\Sigma h^{-1}ef = k_\Sigma$$

is an integer. In the presence of magnetic charges the form f representing the electromagnetic field ceases to be closed. Therefore, the geometric quantization scheme applies only to the part of the space-time free of magnetic charges. Since the flux of the electromagnetic field f through Σ is 4π times the total magnetic charge m_Σ surrounded by Σ, we obtain the *Dirac quantization condition*

$$2em\hbar^{-1} = \text{integer}$$

for the elementary magnetic charge m.

We choose the polarization F leading to the Schrodinger representation. The line bundle L_e over $\mathcal{T}^*Y$ induces a complex line bundle $L_e^\sim$ over Y with a connection $\nabla^\sim$ and a connection invariant Hermitian form $\langle , \rangle^\sim$ such that

$$\text{curvature } \nabla^\sim = -ie\hbar^{-1}f.$$

There is a bijection $\lambda \mapsto \lambda^\sim$ between the space of sections λ of L_e covariantly constant along F and the space of sections $\lambda^\sim$ of $L_e^\sim$. The bundle $\sqrt{\wedge^4 F}$ admits a trivializing section ν_g which is covariantly constant along F. Hence, each section σ in the representation space $\mathcal{H}_e$ is of the form

$$\sigma = \lambda \otimes \nu_g,$$

where the section λ of L_e is covariantly constant along F. The scalar product on $\mathcal{H}_e$ is given by

$$(\lambda_1 \otimes \nu_g | \lambda_2 \otimes \nu_g) = \int_Y \langle\lambda_1^\sim, \lambda_2^\sim\rangle^\sim |\det g|^{\frac{1}{2}}.$$

Hence, the representation space $\mathcal{H}_e$ is isomorphic to the space of square integrable sections of $L_e^\sim$ with the isomorphism

$$\lambda \otimes \nu_g \mapsto \lambda^\sim.$$

Each linear operator A on $\mathcal{H}_e$ induces a linear operator $A^\sim$ on the space of sections of $L_e^\sim$.

The fundamental dynamical variables of physical interest are the position type functions $q = \check{q}\circ\pi$, where $\check{q}$ is a function on Y, the momenta p_ζ associated to vector fields ζ on Y and the mass-squared function N defined by

$$N(x) = g(x,x).$$

The quantization of these dynamical variables yields

$$(\mathcal{Q}q)^{\sim}\lambda^{\sim} = \check{q}\lambda^{\sim}$$

$$(\mathcal{Q}p_{\zeta})^{\sim}\lambda^{\sim} = -i\hbar(\nabla_{\zeta}^{\sim} + \tfrac{1}{2}\,\mathrm{Div}\,\zeta)\lambda^{\sim}$$

$$(\mathcal{Q}N)^{\sim}\lambda^{\sim} = -\hbar^2(\Delta^{\sim} - 1/6\ R)\lambda^{\sim},$$

where $\Delta^{\sim}$ denotes the Laplace-Beltrami operator defined in terms of the metric g on Y and the connection $\nabla^{\sim}$ in $L_e^{\sim}$. The interpretation of N as the square of the mass and the expression for $(\mathcal{Q}N)^{\sim}$ given here shows that the Klein-Gordon equation

$$-\hbar^2(\Delta^{\sim} - 1/6\ R)\lambda^{\sim} = m^2\lambda^{\sim}$$

can be interpreted as the equation for determining the eigenvectors of the mass-squared operator.

The formulation of relativistic dynamics presented here treats the mass and the charge asymmetrically. The mass is a dynamical variable while the charge is a fixed parameter in the theory. This asymmetry disappears in the five-dimensional theory of Kaluza [see Bergmann (1942)]. In the generalized Kaluza theory the configuration space Z is a T^1-principal fibre bundle over the space-time manifold Y, where T^1 is the multiplicative group of complex numbers with absolute value 1. The Lie algebra of T^1 is identified with the real numbers $\mathbb{R}$ by associating to each number r the one-parameter group $t \mapsto \exp(ie_0rt/h)$, where e_0 is a parameter interpreted as the elementary charge. There is a T^1-invariant metric k on Z such that the fundamental vector field η_1 on Z corresponding to the real number 1 has unit length. This structure induces a connection in Z the curvature of which is interpreted as the electromagnetic field f. The horizontal part of k gives rise to a metric g on Y which represents the gravitational field.

The phase space is the cotangent bundle space $\mathscr{T}^*Z$ of Z with the Lagrange bracket given by the canonical form $d\theta_Z$. The canonical momentum p_{η_1} in the direction of η_1 is interpreted as the charge Q. With this interpretation, the Hamiltonian vector field of the pull-back to Z of the mass-squared function N gives rise to the equations of motion of a particle with arbitrary mass and charge moving in the external electromagnetic field f and the gravitational field g.

We study the quantization of this system using the polarization $\hat{F}$ spanned by the Hamiltonian vector field ξ_Q of the charge Q and the Hamiltonian vector fields of the pull-backs to $\mathscr{T}^*Z$ of the position type functions on $\mathscr{T}^*Y$. Since the integral curves of ξ_Q are periodic, we have non-trivial Bohr-Sommerfeld conditions. For each integer n,

$$S_n = Q^{-1}(ne_o)$$

is a connected component of the Bohr-Sommerfeld variety, and we denote by $\mathscr{H}_n$ the subspace of the representation space $\mathscr{H}$ consisting of sections with supports in S_n. The representation space is the direct sum of the $\mathscr{H}_n$,

$$\mathscr{H} = \oplus_{n\in Z} \mathscr{H}_n .$$

According to the general argument leading to the superselection rules, for each function f quantizable in this representation the quantum operator $\mathscr{Q}f$ maps each of the subspaces $\mathscr{H}_n$ to itself. Hence, $\mathscr{Q}f$ is completely specified by the collection of its restrictions $\mathscr{Q}_n f$ to $\mathscr{H}_n$. Each subspace $\mathscr{H}_n$ is the eigenspace of the charge operator corresponding to the eigenvalue

$$e = ne_o.$$

Moreover, there is a unitary isomorphism $\mathscr{U}_n : \mathscr{H}_n \to \mathscr{H}_e$, where $\mathscr{H}_e$ is the representation space corresponding to the charge e introduced before, which intertwines the quantizations of the positions, momenta and the mass-squared function. Hence, the quantization of the Kaluza theory in the representation given by the polarization $\hat{F}$ gives rise to the relativistic quantum dynamics of a particle with arbitrary mass and charge.

1.11. Pauli representation

As the last illustration of the relation between geometric quantization and quantum mechanics, we discuss the Pauli theory of spin. The presentation given here uses the results of Souriau (1970) and Baxter (in preparation).

A classical interpretation of spin is that of an internal angular momentum. Thus, a classical state of a non-relativistic particle with spin $r > 0$ is specified by the position $\underline{q}$, the momentum $\underline{p}$ and the spin vector $\underline{S}$ such that

$$\underline{S}^2 = r^2.$$

In the presence of a magnetic field $\underline{B} = (B_1, B_2, B_3)$, the Lagrange bracket is given by

$$\omega = \sum_i dp_i \wedge dq^i - \tfrac{1}{2} r^{-2} \sum_{ijk} \varepsilon_{ijk} S_i dS_j \wedge dS_k$$
$$+ \tfrac{1}{2} e \sum_{ijk} \varepsilon_{ijk} B_i dq^j \wedge dq^k.$$

The classical dynamics is given by the Hamiltonian vector

field ξ_H of the energy function

$$H = \underline{p}^2/2m + eV(\underline{q}) - (e/mc)\underline{S}\cdot\underline{B}(\underline{q}).$$

A prequantization line bundle L with a connection ∇ satisfying the prequantization condition exists if and only if

$$r = s\hbar,$$

where $2s$ is a positive integer. Thus the prequantization condition leads to the quantization of spin. We choose a polarization F spanned by the Hamiltonian vector fields of $\underline{q}$ and the complex coordinates

$$z_\pm = \frac{S_1 \mp iS_2}{r \pm S_3}\ .$$

A section λ of L is covariantly constant along $\xi_{z_\pm}$ if and only if it is holomorphic in the complex coordinates $z_\pm$. The space of sections holomorphic in $z_\pm$ admits $2s+1$ linearly independent sections $\lambda_0,\ldots,\lambda_{2s}$ such that each section λ of L covariantly constant along F is of the form

$$\lambda = \sum_m \psi_m(\underline{q})\lambda_m,$$

where the $\psi_m(\underline{q})$ are functions of the position variables only.

According to the general principles of geometric quantization, the representation space should consist of sections of $L \otimes \sqrt{\wedge^4 F}$ covariantly constant along F. However, this would give a $(2s+2)$-dimensional space of spin states. In order to circumvent this difficulty, we choose instead the space of sections of $L \otimes \sqrt{\wedge^3 D^{\mathbb{C}}}$ covariantly constant along F to form the representation space $\mathscr{H}$. This choice is justified by the agreement between the quantum theory it engenders and the Pauli theory of spin. It requires certain modifications to the quantization procedure, the theoretical significance of which has yet to be studied.

The bundle $\sqrt{\wedge^3 D^{\mathbb{C}}}$ admits a section $\nu_{\underline{\tilde{\xi}}}$ covariantly constant along F. Hence, each element σ of the representation space $\mathscr{A}$ is of the form

$$\sigma = \sum_m \psi_m(\underline{q})\lambda_m \otimes \nu_{\underline{\tilde{\xi}}}.$$

The scalar product on $\mathscr{A}$ is given by

$$(\sigma|\sigma') = \sum_m \int_{\mathbf{R}^3} \psi_m(\underline{q})\overline{\psi}_m{}'(\underline{q})\,d^3q.$$

The quantization of the position and the momentum variables yields

$$\mathscr{Q}q^i\Big[\sum_n \psi_n(\underline{q})\lambda_n \otimes \nu_{\underline{\tilde{\xi}}}\Big] = \sum_n q^i\psi_n(\underline{q})\lambda_n \otimes \nu_{\underline{\tilde{\xi}}}$$

$$\mathscr{Q}p_j\Big[\sum_n \psi_n(\underline{q})\lambda_n \otimes \nu_{\underline{\tilde{\xi}}}\Big] = \sum_n\Big\{\Big[-i\hbar\frac{\partial}{\partial q^j} - eA_j(\underline{q})\Big]\psi_n(\underline{q})\Big\}\lambda_n \otimes \nu_{\underline{\tilde{\xi}}},$$

where $\underline{A} = (A_1, A_2, A_3)$ is a vector potential for the magnetic field $\underline{B}$:

$$\underline{B} = \operatorname{curl} \underline{A}.$$

The quantized spin operators $\mathscr{Q}s_i$ are given by $(2s+1) \times (2s+1)$ matrices $\underline{D}_i$ with entries $d_{i;m,n}$ such that

$$\mathscr{Q}s_i\Big[\sum_n \psi_n(\underline{q})\lambda_n \otimes \nu_{\underline{\tilde{\xi}}}\Big] = \sum_{mn} d_{i;m,n}\psi_m(\underline{q})\lambda_n \otimes \nu_{\underline{\tilde{\xi}}},$$

and the quantization of the total angular momentum vector $\underline{J}$ yields

$$\mathscr{Q}J_i\Big[\sum_n \psi_n(\underline{q})\lambda_n \otimes \nu_{\underline{\tilde{\xi}}}\Big]$$
$$= \sum_n\Big\{\sum_{jk}\varepsilon_{ijk}q^j\Big[-i\hbar\frac{\partial}{\partial q^k} - eA_k(\underline{q})\Big]\psi_n(\underline{q})\Big\}\lambda_n \otimes \nu_{\underline{\tilde{\xi}}}$$
$$+ \sum_{mn} d_{i;m,n}\psi_m(\underline{q})\lambda_n \otimes \nu_{\underline{\tilde{\xi}}}.$$

Finally, the quantization of the energy H by means of the Blattner-Kostant-Sternberg kernels gives the result

$$\mathscr{Q}H\Big[\sum_n \psi_n\lambda_n \otimes \nu_{\underline{\tilde{\xi}}}\Big] = (-e/mc)\sum_j B_j\, \mathscr{Q}S_j\Big[\sum_n \psi_n\lambda_n \otimes \nu_{\underline{\tilde{\xi}}}\Big]$$

$$+ \sum_n\Big\{\Big[-\frac{\hbar^2}{2m}\sum_{jk}\Big(\frac{\partial}{\partial q^j} - ie\hbar^{-1}A_j\Big)\Big(\frac{\partial}{\partial q^k} - ie\hbar^{-1}A_k\Big) + eV\Big]\psi_n\Big\}\lambda_n \otimes \nu_{\underline{\tilde{\xi}}}.$$

For spin $s = \frac{1}{2}$, the equation for the eigenstates of $\mathscr{Q}H$ corresponds to the Pauli equation, cf. Fermi (1961).

2. HAMILTONIAN DYNAMICS

The phase space description of classical mechanics due to Hamilton is the starting point of the geometric quantization scheme. A brief review of Hamiltonian dynamics, formulated in the language of differential geometry, is given here in order to establish the notation.

2.1. Poisson algebra

The phase space of a dynamical system is represented by an even dimensional manifold X endowed with a *symplectic form* ω defined by the Lagrange bracket. The 2-form ω is closed:

$$d\omega = 0, \tag{2.1}$$

and non-degenerate:

$$\xi \rfloor \omega = 0 \Rightarrow \xi = 0. \tag{2.2}$$

Here, $\rfloor$ denotes the left interior product defined by

$$(\xi \rfloor \omega)(\zeta) = \omega(\xi, \zeta) \tag{2.3}$$

for any two vector fields ξ and ζ on X.

A *canonical transformation* is a ω-preserving diffeomorphism of X onto itself. Infinitesimal diffeomorphisms are given by vector fields; in particular, a vector field ξ on X corresponds to an infinitesimal canonical transformation if and only if the Lie derivative of ω with respect to ξ vanishes:

$$\mathcal{L}_\xi \omega = 0. \tag{2.4}$$

The Lie derivative, exterior derivative, and the left interior product are related by the formula

$$\mathcal{L}_\xi \omega = \xi \rfloor d\omega + d(\xi \rfloor \omega). \tag{2.5}$$

Since ω is closed, it follows that a vector field ξ corresponds to an infinitesimal canonical transformation if and only if $\xi \rfloor \omega$ is closed.

Let f be a smooth function on X. Since ω is nondegenerate there exists a unique vector field ξ_f satisfying

$$\xi_f \rfloor \omega = -df. \tag{2.6}$$

The vector field ξ_f satisfying Eq. (2.6) is called the *Hamiltonian vector field* of f. We denote by $\phi_f^{\,t}$ the local one-parameter group of local diffeomorphisms of X generated by ξ_f. Since the diffeomorphisms generated by Hamiltonian vector fields preserve ω, we refer to $\phi_f^{\,t}$ as the local one-parameter group of local canonical transformations generated by f. If the vector field ξ_f is complete, then $\phi_f^{\,t}$ is a one-parameter group of canonical transformations.

If ξ_f and ξ_g are the Hamiltonian vector fields of f and g, respectively, then their Lie bracket $[\xi_f, \xi_g]$ is the Hamiltonian vector field of $\omega(\xi_f, \xi_g)$,

$$[\xi_f, \xi_g] \rfloor \omega = -d(\omega(\xi_f, \xi_g)). \tag{2.7}$$

The function $-\omega(\xi_f, \xi_g)$ is called the *Poisson bracket* of f and g and will be denoted by $[f,g]$. The following identities are useful in computation:

$$[f,g] = -\omega(\xi_f,\xi_g) = \omega(\xi_g,\xi_f) = -\xi_f g = \xi_g f. \tag{2.8}$$

One can rewrite Eq. (2.7) in terms of the Poisson bracket as follows:

$$[\xi_f,\xi_g] = -\xi_{[f,g]}. \tag{2.9}$$

Since ω is closed, the Poisson bracket satisfies the Jacobi identity

$$[f,[g,k]] + [k,[f,g]] + [g,[k,f]] = 0. \tag{2.10}$$

Hence, the space $C^\infty(X)$ of all smooth functions on (X,ω) is a Lie algebra under the Poisson bracket operation. It is called the *Poisson algebra* of (X,ω).

Equation (2.9) implies that the mapping which associates to each smooth function f its Hamiltonian vector field ξ_f is an antihomomorphism of the Poisson algebra into the Lie algebra of all smooth vector fields on X. The kernel of this antihomomorphism consists of all functions constant on X; we assume X to be connected.

2.2. Local expressions

In most applications the dynamical system under consideration has a physically distinguished configuration space Y, i.e., $X = \mathcal{T}^*Y$. The Lagrange bracket is given by

$$\omega = d\theta_Y, \tag{2.11}$$

where θ_Y is the *canonical 1-form* on $\mathcal{T}^*Y$ defined as follows: for each $x \in \mathcal{T}^*Y$ and each $u \in \mathcal{T}_x\mathcal{T}^*Y$, set

$$\theta_Y(u) = x(\mathcal{T}\pi(u)). \tag{2.12}$$

Here, $\pi: \mathcal{T}^*Y \to Y$ denotes the cotangent bundle projection and

$\mathcal{T}\pi: \mathcal{T}\mathcal{T}^*Y \to \mathcal{T}Y$ is the derived map of π. Using coordinate systems on Y one can construct *canonical* coordinate systems on $\mathcal{T}^*Y$ as follows. If $(U, \check{q}^1, \ldots, \check{q}^n)$ is a coordinate system on Y, then each $x \in \pi^{-1}(U)$ can be written uniquely as

$$x = \sum_{i=1}^{n} p_i(x) d\check{q}^i_{\pi(x)}. \tag{2.13}$$

This gives rise to a coordinate chart $(\pi^{-1}(U), q^1, \ldots, q^n, p_1, \ldots, p_n)$ on $\mathcal{T}^*Y$, where $q^i = \check{q}^i \circ \pi$. With respect to this chart we have

$$\theta_Y = \sum_{i=1}^{n} p_i dq^i \tag{2.14}$$

and

$$\omega = \sum_{i=1}^{n} dp_i \wedge dq^i. \tag{2.15}$$

For each function f on $\mathcal{T}^*Y$, its restriction to $\pi^{-1}(U)$ can be expressed as a function of the coordinates $(q^1, \ldots, q^n, p_1, \ldots, p_n)$ which will be denoted $f(p,q)$. The restriction to $\pi^{-1}(U)$ of the Hamiltonian vector field ξ_f of f can be expressed in terms of the vector fields $\frac{\partial}{\partial q^i}$, $\frac{\partial}{\partial p_i}$, $i = 1, \ldots, n$, as follows:

$$\xi_f | \pi^{-1}(U) = \sum_{i=1}^{n} \left(\frac{\partial f}{\partial p_i} \frac{\partial}{\partial q^i} - \frac{\partial f}{\partial q^i} \frac{\partial}{\partial p_i} \right). \tag{2.16}$$

Note that $\frac{\partial}{\partial q^i}$ is the Hamiltonian vector field of the coordinate function p_i, and $-\frac{\partial}{\partial p_i}$ is the Hamiltonian vector field of q^i. A curve $\gamma: (a,b) \to \mathcal{T}^*Y$, such that $\text{Im}\,\gamma \subset \pi^{-1}(U)$, is an integral curve of ξ_f if and only if it satisfies the canonical equations of Hamilton

$$\frac{d}{dt} q^i(\gamma(t)) = \frac{\partial f}{\partial p_i}(\gamma(t)) \tag{2.17}$$

and

$$\frac{d}{dt} p_i(\gamma(t)) = -\frac{\partial f}{\partial q^i}(\gamma(t)) \tag{2.18}$$

for each $t \in (a,b)$ and each $i = 1,2,\ldots,n$. The Poisson bracket of two functions $f(q,p)$ and $g(q,p)$ is given on $\pi^{-1}(U)$ by the standard expression

$$[f,g](q,p) = \sum_{i=1}^{n} \left(\frac{\partial f}{\partial q^i} \frac{\partial g}{\partial p_i} - \frac{\partial f}{\partial p_i} \frac{\partial g}{\partial q^i} \right). \tag{2.19}$$

2.3. Relativistic charged particle

As an example of the phase space formulation of dynamics, let us consider the relativistic dynamics of a particle with charge e acted on by external gravitational and electromagnetic fields. In this case Y is the space-time manifold. We denote by g the metric on Y, with signature $(-,-,-,+)$, which describes the gravitational field, and by f the closed 2-form on Y characterizing the electromagnetic field. The phase space is the cotangent bundle space of Y, $X = \mathcal{T}^*Y$, and the Lagrange bracket is

$$\omega_e = d\theta_Y + e\pi^* f, \tag{2.20}$$

where e is the charge of the particle and $\pi^* f$ is the pull-back of the 2-form f to $\mathcal{T}^*Y$. The incorporation of the electromagnetic field strength into the expression for the Lagrange bracket enables one to discuss the Hamiltonian dynamics of a relativistic charged particle without any reference to the electromagnetic potentials.

The metric g on Y defines a vector bundle isomorphism $\mathcal{T}Y \to \mathcal{T}^*Y$ and induces a symmetric bilinear form on $\mathcal{T}^*Y$ which we also denote by g. Let N be the function on $\mathcal{T}^*Y$ defined by

$$N(x) = g(x,x) \tag{2.21}$$

for each $x \in \mathscr{T}^*Y$. If $N(x) > 0$, then $M(x) = \sqrt{N(x)}$ is the mass of the particle in the classical state x. The dynamics of a massive particle is given by the Hamiltonian vector field ξ_M of M. However, from the computational point of view, it is more convenient to work with the Hamiltonian vector field ξ_N of N which is parallel to ξ_M. It is known in differential geometry that the Hamiltonian vector field of N, defined with respect to the symplectic form $d\theta_Y$, yields the equations of the geodesics in X. The correction $e\pi^*f$ in the symplectic form ω_e gives rise to the Lorentz force.

We demonstrate this by direct computations in terms of the local chart $(\pi^{-1}(U), q^1, \ldots, q^4, p_1, \ldots, p_4)$ on $\mathscr{T}^*Y$ induced by the chart $(U, \check{q}^1, \ldots, \check{q}^4)$ on Y. The metric g defines a matrix-valued function (g_{ij}) on U such that

$$g|U = \sum_{ij} g_{ij} d\check{q}^i \otimes d\check{q}^j. \tag{2.22}$$

Similarly, the restriction of f to U can be expressed in terms of its components f_{ij} as follows:

$$f|U = \tfrac{1}{2} \sum_{ij} f_{ij} d\check{q}^i \wedge d\check{q}^j. \tag{2.23}$$

Therefore

$$N|\pi^{-1}(U) = \sum_{ij} (g^{ij} \circ \pi) p_i p_j, \tag{2.24}$$

where (g^{ij}) is the inverse of the matrix (g_{ij}), and

$$\omega_e|\pi^{-1}(U) = \sum_i dp_i \wedge dq^i + e/2 \sum_{ij} (f_{ij} \circ \pi) dq^i \wedge dq^j. \tag{2.25}$$

Equations (2.24) and (2.25) yield the following local expression for the Hamiltonian vector field ξ_N of N:

$$\begin{aligned} \xi_N|\pi^{-1}(U) &= 2 \sum_{ij} (g^{ij} \circ \pi) p_j \frac{\partial}{\partial q^i} \\ &\quad - \sum_{ijk} \Big[(g^{ij}{}_{,k} \circ \pi) p_i p_j + 2e (f_{ik} \circ \pi)(g^{ij} \circ \pi) p_j \Big] \frac{\partial}{\partial p_k}, \end{aligned} \tag{2.26}$$

where

$$g^{ij}{}_{,k} := \frac{\partial g^{ij}}{\partial \check{q}^k} . \tag{2.27}$$

The orbits of the one-parameter group $\phi_N{}^t$ of diffeomorphisms of $\mathscr{T}^*Y$ generated by ξ_N satisfy the equations

$$\frac{d}{dt}(q^i \circ \phi_N{}^t) = 2 \sum_j [(g^{ij} \circ \pi) p_j] \circ \phi_N{}^t \tag{2.28}$$

and

$$\frac{d}{dt}(p_i \circ \phi_N{}^t) = - \sum_{jk} [(g^{jk}{}_{,i} \circ \pi) p_j p_k + \\ + 2e(f_{ki} \circ \pi)(g^{kj} \circ \pi) p_j] \circ \phi_N{}^t . \tag{2.29}$$

Introducing the Christoffel symbols of the metric connection in Y defined by

$$\Gamma_{mn}{}^i = \tfrac{1}{2} \sum_j g^{ij}(g_{jm,n} + g_{jn,m} - g_{mn,j}), \tag{2.30}$$

where

$$g_{mn,j} := \frac{\partial g_{mn}}{\partial \check{q}^j} , \tag{2.31}$$

we can rewrite Eqs. (2.28) and (2.29) as a second-order equation

$$\frac{d^2}{dt^2}(q^i \circ \phi_N{}^t) + \sum_{mn} (\Gamma_{mn}{}^i \circ \pi \circ \phi_N{}^t) \frac{d}{dt}(q^m \circ \phi_N{}^t) \frac{d}{dt}(q^n \circ \phi_N{}^t) \\ = 2e \sum_{jk} [(g^{ij} f_{jk}) \circ \pi \circ \phi_N{}^t] \frac{d}{dt}(q^k \circ \phi_N{}^t) . \tag{2.32}$$

Since

$$\xi_M = (2M)^{-1} \xi_N \tag{2.33}$$

for $N > 0$, the parameter s along the integral curves of ξ_M is related to t by

$$ds = 2M dt . \tag{2.34}$$

If $\gamma(s)$ is the projection onto Y of an integral curve of ξ_M, Eqs. (2.23) and (2.34) yield

$$M\left[\frac{d^2}{ds^2}\check{q}^i(\gamma(s)) + \sum_{mn}\Gamma_{mn}{}^i(\gamma(s))\frac{d}{ds}\check{q}^m(\gamma(s))\frac{d}{ds}\check{q}^n(\gamma(s))\right]$$
$$= e\sum_{jk} g^{ij}(\gamma(s))f_{jk}(\gamma(s))\frac{d}{ds}\check{q}^k(\gamma(s)). \tag{2.35}$$

The left hand side of Eq. (2.35) is the absolute derivative of the tangent vector to $\gamma(s)$ multiplied by the mass M (the mass is constant along $\gamma(s)$), while the right hand side is the Lorentz force. The dynamics of the relativistic charged particle described here will be geometrically quantized in Chapter 10.

2.4. Non-relativistic dynamics

In non-relativistic dynamics the phase space is defined in terms of a chosen inertial frame. The absolute time of Newtonian dynamics can be represented by a one-dimensional affine space T endowed with a translation invariant 1-form dt corresponding to a fixed time scale. To each $t \in T$, there is associated the phase space (X_t, ω_t) at the time t. The collection

$$Z = \bigcup_{t\in T} X_t \tag{2.36}$$

of all the phase spaces at all times t is called the *evolution space* of the system. Let $\tau: Z \to T$ denote the projection map defined by $\tau^{-1}(t) = X_t$ for each $t \in T$. An inertial frame defines a trivialization of τ which induces symplectic isomorphisms of the fibres with a typical fibre (X, ω). It is this typical fibre which is used in the Hamiltonian formulation of non-relativistic dynamics. In order to have a clear understanding of the set-up described here, we discuss below the non-relativistic dynamics of a single particle.

The Galilean space-time, which we denote also by Y, is a 4-dimensional affine space. The notion of simultaneity defines an affine mapping $\tau\colon Y \to T$. Each fibre of τ is a 3-dimensional Euclidean space, and the translations of Y induce isometries of the fibres of τ.

For each $t \in T$, the fibre $Y_t = \tau^{-1}(t)$ represents the physical space at time t. The phase space at time t is the cotangent bundle space of Y_t,

$$X_t = \mathscr{T}^*Y_t, \tag{2.37}$$

and the Lagrange bracket on X_t is given by

$$\omega_t = d\theta_{Y_t}. \tag{2.38}$$

The Newtonian dynamics of a single particle is determined by an extension of the collection $\{\omega_t \mid t \in T\}$ of forms on the fibres of $\tau\colon Z \to T$ to a 2-form Ω on Z such that $\omega_t = \Omega|X_t$. Since this description of dynamics can be put into Hamiltonian form if and only if Ω is closed, we assume that

$$d\Omega = 0, \tag{2.39}$$

Let N denote the characteristic distribution of Ω,

$$N = \{v \in \mathscr{T}Z \,|\, v \,\lrcorner\, \Omega = 0\}. \tag{2.40}$$

The distribution N is one-dimensional and involutive. For each $z \in Z$, the evolution of the classical state z is given by the integral manifold of N through z. Let ζ be the vector field spanning N normalized by

$$\tau^*dt(\zeta) = 1. \tag{2.41}$$

The vector field ζ is complete and we denote by ϕ^s the one-parameter group of diffeomorphisms of Z generated by ζ.

Since $\zeta \rfloor \Omega = 0$ and Ω is closed, it follows that the diffeomorphisms ϕ^s preserve Ω. The normalization of ζ given by Eq. (2.41) implies that each ϕ^s preserves the fibre bundle structure of Z given by $\tau: Z \to T$ and that ϕ^s induces a translation of T denoted by $t \mapsto t+s$. For each $t \in T$, we denote by $\phi_t{}^s: \mathscr{T}^*Y_t \to \mathscr{T}^*Y_{t+s}$ the diffeomorphism obtained by restricting ϕ^s to $\mathscr{T}^*Y_t$. Since ϕ^s preserves Ω it follows that $\phi_t{}^{s*}\omega_{t+s} = \omega_t$. Thus, the non-relativistic dynamics of a particle can be represented by a family of canonical transformations $\phi_t{}^s$ between the phase spaces $(\mathscr{T}^*Y_t, \omega_t)$ and $(\mathscr{T}^*Y_{t+s}, \omega_{t+s})$ which satisfy the composition law

$$\phi_{t+s}{}^{s'} \circ \phi_t{}^s = \phi_t{}^{s+s'}. \tag{2.42}$$

A canonical transformation $\phi_t{}^s: \mathscr{T}^*Y_t \to \mathscr{T}^*Y_{t+s}$ can be described in terms of a generating function $S_t{}^s: Y_t \times Y_{t+s} \to R$ provided, for each $y \in Y_t$ and $y' \in Y_{t+s}$, there exist $z \in \mathscr{T}^*Y_t$ and $z' \in \mathscr{T}^*Y_{t+s}$ projecting onto y and y', respectively, such that $z' = \phi_t{}^s(z)$. This condition is usually satisfied in non-relativistic dynamics, at least for sufficiently small values of the parameter s. To determine $S_t{}^s$, note that the assumptions that Ω is closed and Z is contractible imply the existence of a 1-form Θ such that

$$\Omega = d\Theta. \tag{2.43}$$

Moreover, we can choose Θ so that, for each $t' \in T$,

$$\Theta \,|\, \mathscr{T}^*Y_{t'} = \theta_{Y_{t'}}. \tag{2.44}$$

This restriction defines Θ up to a differential of a function on T.

For each $y \in Y_t$ and $y' \in Y_{t+s}$, we set

$$S_t^{\,s}(y,y') = \int_{\mathrm{Im}\,\gamma} \Theta, \tag{2.45}$$

where γ is the integral curve of ζ such that $z = \gamma(0)$ projects onto y and $z' = \gamma(s)$ projects onto y'. It can be shown that the function $S_t^{\,s}$ is well-defined by Eq. (2.45) and that

$$dS_t^{\,s} \mid (\mathscr{T}_y^*Y_t \oplus \mathscr{T}_{y'}^*Y_{t+s}) = (-z) \oplus z'. \tag{2.46}$$

An *inertial frame* is given by specifying a product structure of the space-time, $Y \approx E \times T$, where E is a 3-dimensional Euclidean space. The inertial frame induces a product structure of the evolution space, $Z \approx \mathscr{T}^*E \times T$, with the projections $\upsilon\colon Z \to \mathscr{T}^*E$ and $\tau\colon Z \to T$. The symplectic manifold $(\mathscr{T}^*E, d\theta_E)$ is the phase space defined by the chosen inertial frame.

The pull-back $\upsilon^*\theta_E$ of θ_E is a 1-form on Z such that, for each $t \in T$,

$$\upsilon^*\theta_E \mid \mathscr{T}^*Y_t = \theta_{Y_t}. \tag{2.47}$$

Eqs. (2.47) and (2.44) imply that Θ and $\upsilon^*\theta_E$ differ by a 1-form vanishing on the fibres of $\tau\colon Z \to T$. Hence, there exists a unique function $\tilde{H}\colon Z \to \mathbb{R}$ such that

$$\Theta = \upsilon^*\theta_E - \tilde{H}\,dt. \tag{2.48}$$

The function $\tilde{H}$ is the (time-dependent) Hamiltonian, relative to the chosen inertial frame, of the dynamics described by Θ. Since Θ is determined by Eqs. (2.43) and (2.44) up to a differential of a function on T, it follows that $\tilde{H}$ is determined by the dynamics Ω and the chosen inertial frame up to an arbitrary smooth function on T.

If the Hamiltonian $\tilde{H}$ is constant along the fibres of $\upsilon\colon Z \to \mathscr{T}^*E$, it projects to a function $H\colon \mathscr{T}^*E \to \mathbb{R}$ such that

$\tilde{H} = H \circ \upsilon$. In this case, the projections to $\mathscr{T}^*E$ of the integral curves of ζ in Z can be represented by the integral curves of the Hamiltonian vector field ξ_H of H in the phase space $(\mathscr{T}^*E, d\theta_E)$ defined by the inertial frame.

The choice of an initial time yields an affine isomorphism $T \to \mathbb{R}$ such that, if we denote by $t: Z \to \mathbb{R}$ its pull-back to Z, the 1-form dt corresponding to the choice of a time scale coincides with the differential of the function t. Let q^1, q^2, q^3 denote the pull-backs to Z of some Cartesian coordinates on E, and p_1, p_2, p_3 denote the conjugate momentum functions. Then,

$$\upsilon^* \theta_E = \sum_i p_i dq^i \tag{2.49}$$

$$\Theta = \sum_i p_i dq^i - \tilde{H}\, dt \tag{2.50}$$

and

$$\Omega = \sum_i dp_i \wedge dq^i - d\tilde{H} \wedge dt. \tag{2.51}$$

In the coordinate system on Z given by the functions p_i, q^i and t, the vector field ζ can be written

$$\zeta = \sum_i \left(\frac{\partial \tilde{H}}{\partial p_i} \frac{\partial}{\partial q^i} - \frac{\partial \tilde{H}}{\partial q^i} \frac{\partial}{\partial p_i} \right) + \frac{\partial}{\partial t} . \tag{2.52}$$

Thus, the integral curves of ζ satisfy the usual Hamiltonian equations of motion with the time-dependent Hamiltonian $\tilde{H}$.

The generating function $S_t{}^s$ defined by Eq. (2.45) can be expressed in terms of a function S depending on the parameters t and $t+s$, the coordinates $\underline{q} = (q^1, q^2, q^3)$ of a point $y \in Y_t$, and the coordinates $\underline{q}' = (q^{1'}, q^{2'}, q^{3'})$ of a point $y' \in Y_{t+s}$:

$$S_t{}^s(y, y') = S(t, \underline{q};\ t+s,\ \underline{q}'). \tag{2.53}$$

Let $\underline{p} = (p_1, p_2, p_3)$ denote the momentum coordinates of $z \in \mathcal{T}_y{}^*Y_t$ and $\underline{p}' = (p'_1, p'_2, p'_3)$ the momentum coordinates of $z' \in \mathcal{T}_{y'}{}^*Y_{t+s}$, where $z' = \phi_t{}^s(z)$. We can rewrite Eq. (2.46) in the familiar form

$$\frac{\partial}{\partial q^i} S(t,\underline{q};\ t+s,\underline{q}') = -p_i, \quad \frac{\partial}{\partial q'^i} S(t,\underline{q};\ t+s,\underline{q}') = p_i'. \qquad (2.54)$$

The Hamiltonian formalism for non-relativistic dynamics presented here extends, with obvious modifications, to the dynamics of a system of particles subject to time-independent holonomic constraints, e.g., a rigid body. The modifications consist of the interpretation of Y_t as the configuration manifold at time t, so that $Y = \bigcup_{t \in T} Y_t$ represents the configuration space-time. An inertial frame induces a product structure of Y, $Y \approx W \times T$, where W is the configuration manifold defined by the inertial frame.

3. PREQUANTIZATION

The aim of this chapter is to construct a linear monomorphism from the Poisson algebra of (X,ω) to the space of linear operators on an appropriate Hilbert space, associating to each function f on X a linear operator $\mathscr{P}f$, so that the commutation relations

$$[\mathscr{P}f, \mathscr{P}g] = i\hbar\mathscr{P}[f,g] \tag{3.1}$$

are satisfied for each pair of functions f and g on X. It should be noted that the mapping $f \mapsto (\hbar/i)\xi_f$ satisfies (3.1) but it fails to be a monomorphism since its kernel consists of the space of all constant functions on X. Thus, we need a central extension of the Lie algebra of Hamiltonian vector fields on X by the additive group $\mathbb{R}$ of real numbers.

3.1. Connections in line bundles

We begin with a review of some basic results in the theory of connections in complex line bundles. In the following $\mathbb{C}^\times$ denotes the multiplicative group of nonzero complex numbers. The Lie algebra of $\mathbb{C}^\times$ is identified with $\mathbb{C}$ by associating to each complex number c the one-parameter subgroup $e^{2\pi ict}$ of $\mathbf{C}^\times$.

Let L be a complex line bundle over X and $L^\times$ the bundle obtained from L by removing the zero section. $L^\times$ is the $\mathbf{C}^\times$-principal fibre bundle over X associated to L. The fact that $L^\times$ is contained in L substantially simplifies the presentation. We denote by π the fibre bundle projections $L \to X$ and $L^\times \to X$.

The space of sections λ of the line bundle L is isomorphic to the space of complex-valued functions $\lambda^\#$ on $L^\times$ satisfying the conditions

$$\lambda^\#(cz) = c^{-1}\lambda^\#(z) \tag{3.2}$$

for each $c \in \mathbf{C}^\times$ and each $z \in L^\times$; the isomorphism $\lambda^\# \mapsto \lambda$ being given by

$$\lambda(\pi(z)) = \lambda^\#(z)z \tag{3.3}$$

for each $z \in L^\times$. If, for some $x \in X$, $\lambda(x) \neq 0$, then $\lambda(x) \in L^\times$ and (3.3) implies that $\lambda(x) = \lambda^\#(\lambda(x))\lambda(x)$. Hence

$$\lambda^\# \circ \lambda = 1 \tag{3.4}$$

for each non-vanishing local section λ of L. For every $c \in \mathbf{C}$, we denote by η_c the *fundamental vector field* on $L^\times$ corresponding to c. That is, for each $z \in L^\times$, $\eta_c(z)$ is the tangent vector at z of the curve $t \mapsto e^{2\pi ict}z$. Hence, for each complex-valued function f on $L^\times$,

$$(\eta_c f)(z) = \frac{d}{dt} f(e^{2\pi ict}\, z)\Big|_{t=0}. \tag{3.5}$$

For each function $\lambda^\#$ satisfying (3.2), we can compute the action of η_c on $\lambda^\#$ explicitly. We have

$$\begin{aligned}(\eta_c\lambda^\#)(z) &= \frac{d}{dt}\lambda^\#(e^{2\pi ict}z)\Big|_{t=0}\\ &= \frac{d}{dt} e^{-2\pi ict}\lambda^\#(z)\Big|_{t=0} = -2\pi ic\lambda^\#(z).\end{aligned}$$

Hence,

$$\eta_c \lambda^{\#} = -2\pi i c \lambda^{\#}. \tag{3.6}$$

For each function f on X we define a vector field η_f on $L^{\times}$ by

$$\eta_f(z) = \eta_{f(\pi(z))}(z) \tag{3.7}$$

for each $z \in L^{\times}$. Then,

$$\eta_f \lambda^{\#} = -2\pi i (f \circ \pi) \lambda^{\#}. \tag{3.8}$$

Let α be a *connection form* on $L^{\times}$, that is, a $C^{\times}$-invariant complex-valued 1-form α on $L^{\times}$ such that

$$\alpha(\eta_c) = c \tag{3.9}$$

for each $c \in C$. The connection form α defines the *horizontal distribution* hor $\mathscr{T}L$ in $L^{\times}$ given by

$$\text{hor}\,\mathscr{T}L^{\times} = \{u \in \mathscr{T}L^{\times} \mid \alpha(u) = 0\}. \tag{3.10}$$

The *vertical distribution* ver $\mathscr{T}L^{\times}$ in $\mathscr{T}L^{\times}$ consists of all vectors tangent to the fibres of $\pi: L^{\times} \to X$, and $\mathscr{T}L^{\times}$ is the direct sum of the horizontal and the vertical distributions

$$\mathscr{T}L^{\times} = \text{hor}\,\mathscr{T}L^{\times} \oplus \text{ver}\,\mathscr{T}L^{\times}. \tag{3.11}$$

For each vector field ζ on $L^{\times}$ we denote by hor ζ and ver ζ its horizontal and vertical components, respectively. Thus for each $z \in L^{\times}$

$$\zeta(z) = \text{hor}\,\zeta(z) + \text{ver}\,\zeta(z), \tag{3.12}$$

where hor $\zeta(z) \in \text{hor}\,\mathscr{T}_z L^{\times}$ and ver $\zeta(z) \in \text{ver}\,\mathscr{T}_z L^{\times}$. A vector field on $L^{\times}$ is called a *horizontal (vertical)* vector field if its vertical (horizontal) component vanishes identically. The *horizontal lift* of a vector field ξ on X is the unique horizontal vector field $\xi^{\#}$ on $L^{\times}$ projecting onto ξ, i.e., the unique vector field $\xi^\#$ satisfying

$$\alpha(\xi^\#) = 0 \quad \text{and} \quad \mathscr{T}\pi(\xi^\#(z)) = \xi(\pi(z)) \tag{3.13}$$

for each $z \in L^\times$.

The *covariant derivative* of a section λ of L in the direction of a vector field ξ on X, defined by the connection in $L^\times$, is the section $\nabla_\xi\lambda$ of L given by

$$\nabla_\xi\lambda(\pi(z)) = (\xi^\#\lambda^\#)(z)z \tag{3.14}$$

for each $z \in L^\times$, where $\xi^\#$ is the horizontal lift of ξ and $\lambda^\#$ is the function on $L^\times$ satisfying (3.2) and (3.3). Clearly, $\nabla_\xi\lambda$ depends linearly on ξ and on λ, and

$$\nabla_{f\xi}\lambda = f\nabla_\xi\lambda \tag{3.15}$$

$$\nabla_\xi(f\lambda) = (\xi f)\lambda + f\nabla_\xi\lambda \tag{3.16}$$

for each function f on X.

The covariant derivative of a non-vanishing section λ can be related directly to the connection form α as follows. Since λ is a map from X to $L^\times$ such that $\pi \circ \lambda = \text{identity}$, it follows that for each $x \in X$, $\mathscr{T}\lambda(\xi(x))$ is a vector in $\mathscr{T}_{\lambda(x)}L^\times$ and consequently can be decomposed into horizontal and vertical components

$$\begin{aligned}\mathscr{T}\lambda(\xi(x)) &= \text{hor}\,\mathscr{T}\lambda(\xi(x)) + \text{ver}\,\mathscr{T}\lambda(\xi(x)) \\ &= \xi^\#(\lambda(x)) + \eta_{\alpha(\mathscr{T}\lambda(\xi(x)))}(\lambda(x)).\end{aligned}$$

Hence,

$$\begin{aligned}(\xi^\#\lambda^\#)(\lambda(x)) &= d\lambda^\#(\xi^\#(\lambda(x))) \\ &= d\lambda^\#(\mathscr{T}\lambda(\xi(x))) - d\lambda^\#(\eta_{\alpha(\mathscr{T}\lambda(\xi(x)))}(\lambda(x))).\end{aligned}$$

But, $d\lambda^\#(\mathscr{T}\lambda(\xi(x))) = (\xi(\lambda^\#\circ\lambda))(x)$ which vanishes by (3.4), and

$$d\lambda^{\#}(\eta_{\alpha(\mathscr{T}\lambda(\xi(x)))}(\lambda(x)) = (\eta_{\alpha(\mathscr{T}\lambda(\xi(x)))}\lambda^{\#})(\lambda(x))$$
$$= -2\pi i\alpha(\mathscr{T}\lambda(\xi(x)))\lambda^{\#}(\lambda(x))$$
$$= -2\pi i\lambda^{*}\alpha(\xi(x)).$$

Therefore, $(\xi^{\#}\lambda^{\#})(\lambda(x)) = 2\pi i\ \lambda^{*}\alpha(\xi(x))$ which implies that $\nabla_{\xi}\lambda(x) = 2\pi i\lambda^{*}\alpha(\xi(x))\lambda(x)$. Hence, for each non-vanishing section λ of L,

$$\nabla_{\xi}\lambda = 2\pi i\lambda^{*}\alpha(\xi)\lambda. \tag{3.17}$$

For any two vector fields ξ and ξ' on X and any non-vanishing section λ of L we have

$$(\nabla_{\xi}\nabla_{\xi'} - \nabla_{\xi'}\nabla_{\xi} - \nabla_{[\xi,\xi']})\lambda = 2\pi i\lambda^{*}d\alpha(\xi,\xi')\lambda. \tag{3.18}$$

The 2-form $d\alpha$ is the *curvature form* of the connection in $L^{\times}$ defined by α. Since $\alpha(\eta_c) = c$, the curvature form satisfies

$$\eta_c \lrcorner\, d\alpha = 0 \quad \text{and} \quad \mathscr{L}_{\eta_c} d\alpha = 0 \tag{3.19}$$

for each $c \in \mathbf{C}$. Hence $d\alpha$ is the pull-back by π of a closed complex-valued 2-form on X.

3.2. Prequantization line bundle

A *prequantization line bundle* of a symplectic manifold (X,ω) is a complex line bundle L over X with a connection ∇ such that the connection form α satisfies the *prequantization condition*

$$d\alpha = -h^{-1}\pi^{*}\omega, \tag{3.20}$$

where h is Planck's constant. Such a line bundle exists if and only if $h^{-1}\omega$ defines an integral de Rham cohomology class and, if this condition is satisfied, the set of equivalence classes of such line bundles with connection can be parametrized by the group of all unitary characters of the fundamental group of X. Further, we shall assume that there exists a ∇-invariant Hermitian structure $<,>$ on L, where ∇-invariance of $<,>$ means that, for each pair of sections λ and λ' of L and each vector field ξ on X,

$$\xi<\lambda,\lambda'> = <\nabla_\xi\lambda,\lambda'> + <\lambda,\nabla_\xi\lambda'>. \tag{3.21}$$

A ∇-invariant Hermitian structure on a line bundle with a connection form α exists if and only if

$$2\pi i(\alpha - \overline{\alpha}) \quad \text{is exact,} \tag{3.22}$$

and it is determined by α up to a multiplicative positive constant.

Let ζ be a real vector field on $L^\times$ preserving α, that is,

$$\zeta \rfloor d\alpha + d(\alpha(\zeta)) = \mathcal{L}_\zeta\alpha = 0. \tag{3.23}$$

Evaluating (3.23) on η_c and taking into account the fact that $\eta_c \rfloor d\alpha = 0$, we find that

$$\eta_c(\alpha(\zeta)) = 0 \tag{3.24}$$

for all $c \in \mathbf{C}$. Hence, $\alpha(\zeta)$ is constant along the fibres of $\pi: L^\times \to X$ and consequently there exists a function f on X, possibly complex-valued, such that

$$\alpha(\zeta) = -h^{-1} f \circ \pi. \tag{3.25}$$

Here, a factor involving Planck's constant h has been introduced in analogy with the quantization condition (3.20). Eq. (3.25) determines the vertical part of ζ giving

$$\text{ver } \zeta = -\eta_{f/h}, \tag{3.26}$$

where the right hand side is defined by Eq. (3.7). Substituting (3.25) back into (3.23) one gets

$$\text{hor } \zeta \lrcorner \, \pi^*\omega = -d(f \circ \pi). \tag{3.27}$$

Hence, f is a real-valued function and hor ζ is the horizontal lift of the Hamiltonian vector field ξ_f of f,

$$\text{hor } \zeta = \xi_f^{\#}. \tag{3.28}$$

We denote by ζ_f the vector field determined by (3.26) and (3.28), i.e.,

$$\zeta_f = \xi_f^{\#} - \eta_{f/h}. \tag{3.29}$$

The association $f \mapsto \zeta_f$ is a linear isomorphism of the Poisson algebra of (X,ω) onto the Lie algebra of real connection-preserving vector fields on $L^{\times}$. Each vector field ζ_f is $C^{\times}$-invariant and therefore we can define its action on the space of functions $\lambda^{\#}$ satisfying $\lambda^{\#}(cz) = c^{-1}z$, and hence its action on the space of sections of L. Taking into account (3.8) and (3.14) we obtain, for each section λ of L,

$$\zeta_f \lambda^{\#} = [\nabla_{\xi_f} \lambda + (i/\hbar) f\lambda]^{\#} \tag{3.30}$$

where $\hbar = h/2\pi$.

3.3. Prequantization map

Let f be a function on X such that its Hamiltonian vector field ξ_f is complete. Then f generates a one-parameter group ϕ_f^t of canonical transformations of (X,ω). The group ϕ_f^t induces a one-parameter group $\phi_f^{\#t}$ of connection-preserving diffeomorphisms of $L^{\times}$ such that, for each $t \in R$,

$$\pi \circ \phi_f^{\#t} = \phi_f{}^t \circ \pi. \tag{3.31}$$

The group $\phi_f^{\#t}$ is generated by the connection-preserving vector field ζ_f defined by Eq. (3.29). Since each $\phi_f^{\#t}$ preserves the connection form α, it commutes with the action of $\mathbf{C}^\times$ on $L^\times$. Hence, for each $t \in \mathbf{R}$ and each section λ of L, the function $\lambda^\# \circ \phi_f^{\#-t}$ defines a section of L which we denote by $\phi_f{}^t\lambda$,

$$(\phi_f{}^t\lambda)^\# = \lambda^\# \circ \phi_f^{\#-t}. \tag{3.32}$$

The mappings

$$\lambda^\# \mapsto \lambda^\# \circ \phi_f^{\#-t} \tag{3.33}$$

form a one-parameter group of linear transformations on the space of functions on $L^\times$ satisfying Eq. (3.3). Therefore, the mappings

$$\lambda \mapsto \phi_f{}^t\lambda \tag{3.34}$$

form a one-parameter group of linear transformations on the space of sections of L. The *prequantized operator* $\mathscr{P}f$ corresponding to f is defined to be the generator of this one-parameter group:

$$\mathscr{P}f[\lambda] := i\hbar \frac{d}{dt}(\phi_f{}^t\lambda)\Big|_{t=0}. \tag{3.35}$$

The operator $\mathscr{P}f$ can be expressed directly in terms of ζ_f. Since ζ_f generates $\phi_f^{\#t}$, Eq. (3.32) yields

$$\frac{d}{dt}(\phi_f{}^t\lambda)^\# = -\zeta_f(\phi_f{}^t\lambda)^\#. \tag{3.36}$$

Hence,

$$(\mathscr{P}f[\lambda])^\# = -i\hbar\zeta_f\lambda^\#. \tag{3.37}$$

Eq. (3.37) could be used as an alternative definition of $\mathscr{P}f$. In particular, it can be used to define $\mathscr{P}f$ when the vector field ζ_f is incomplete and $\phi_f{}^t\lambda$ need not be defined.

Taking into account Eq. (3.30) we have

$$\mathcal{P}f[\lambda] = [-i\hbar\nabla_{\xi_f} + f]\lambda. \tag{3.38}$$

If f is a constant function, then $\xi_f = 0$, and $\mathcal{P}f$ is the operator of multiplication by f. Thus, the map $f \mapsto \mathcal{P}f$ is a linear monomorphism of the Poisson algebra of (X,ω) into the algebra of differential operators on the space of sections of L. Using the commutation relations for the covariant derivatives, Eq. (3.18), and the properties of the Poisson bracket one can easily verify that Eq. (3.1) is satisfied.

4. REPRESENTATION SPACE

4.1. Polarization

In quantum mechanics one can represent the Hilbert space of states as the space of square integrable complex functions on the spectrum of any given complete set of commuting observables. In the process of quantization, however, one has only the classical phase space (X,ω) to work with, and one has to find a suitable classical counterpart of the notion of a complete set of commuting observables. A natural choice is a set of $n = \frac{1}{2}\dim X$ functions $f_1,\ldots,f_n$ on X, independent at all points of X, satisfying

$$[f_i,f_j] = 0, \quad i,j \in \{1,2,\ldots,n\} \tag{4.1}$$

such that their Hamiltonian vector fields $\xi_{f_1},\ldots,\xi_{f_n}$ are complete. However, for many phase spaces of interest there does not exist such a set. If one drops the assumption that the f_i be real and globally defined, one is led to the notion of a "polarization" of (X,ω). Note that the Hamiltonian vector fields ξ_{f_i}, $i \in \{1,\ldots,n\}$ span over $\mathbb{C}$ an involutive distribution F on X such that

$$\dim_{\mathbb{C}} F = \tfrac{1}{2}\dim X, \tag{4.2}$$

and also that ω restricted to F vanishes identically,

$$\omega | F \times F = 0. \tag{4.3}$$

A complex distribution F satisfying (4.2) and (4.3) is called a *complex Lagrangian distribution* on (X,ω).

A *polarization* of a symplectic manifold (X,ω) is a complex involutive Lagrangian distribution F on X such that $\dim\{F_x \cap \overline{F}_x\}$ is constant, where $\overline{F}$ denotes the complex conjugate of F. The complex distributions $F \cap \overline{F}$ and $F + \overline{F}$ defined by a polarization F are the complexifications of certain real distributions which we denote by D and E respectively:

$$F \cap \overline{F} = D^{\mathbb{C}} \quad \text{and} \quad F + \overline{F} = E^{\mathbb{C}}. \tag{4.4}$$

For each $x \in X$, the vector spaces D_x and E_x are related to each other as follows:

$$E_x = \{u \in \mathscr{T}_x X \mid \omega(u,v) = 0, \text{ for all } v \in D_x\} \tag{4.5}$$

and

$$D_x = \{v \in \mathscr{T}_x X \mid \omega(u,v) = 0, \text{ for all } u \in E_x\}. \tag{4.6}$$

The involutivity of F implies that D is an involutive distribution so that D defines a foliation of X. We denote by X/D the space of all integral manifolds of D and by $\pi_D\colon X \to X/D$ the canonical projection. A polarization F is said to be *strongly admissible* if E is an involutive distribution, the spaces X/D and X/E of the integral manifolds of D and E, respectively, are quotient manifolds of X and the canonical projection $X/D \to X/E$ is a submersion. For a strongly admissible polarization we denote by $\pi_E\colon X \to X/E$ and $\pi_{ED}\colon X/D \to X/E$ the canonical projections.

Let F be a strongly admissible polarization. Then, for each integral manifold Λ of D, the tangent bundle $\mathscr{T}\Lambda$ of Λ is globally spanned by commuting vector fields. This defines a global parallelism in Λ in which the parallel vector fields are the restrictions to Λ of the Hamiltonian vector fields in D. Furthermore, each fibre M of π_{ED} has a Kähler structure such that $F|\pi_D^{-1}(M)$ projects onto the distribution of anti-holomorphic vectors on M. In the remainder of this section we outline proofs of these statements.

Let f be a function on X such that $uf = 0$ for each $u \in E$. Then $(\xi_f \rfloor \omega)(u) = -df(u) = -uf = 0$, and Eq. (4.6) implies that $\xi_f(x) \in D_x$ for each $x \in X$. Therefore, the Hamiltonian vector fields of functions constant along E are in D, and they commute since the Poisson bracket of functions f and g constant along E vanishes: $[f,g] = -\xi_f g = 0$ for $\xi_f(x) \in D_x \subset E_x$. Conversely, if ξ_f is the Hamiltonian vector field of a function f such that $\xi_f(x) \in D_x$ for each $x \in X$, then f is constant along E. Given an integral manifold Λ of D, let x be any point of Λ and $(V;\check{q}^1,\ldots,\check{q}^d)$, $d = \dim D = \operatorname{codim} E$, a chart on X/E at $\pi_E(x)$. The Hamiltonian vector fields $\xi_{q^1},\ldots,\xi_{q^d}$ of $q^1 = \check{q}^1 \circ \pi_E, \ldots,$ $q^d = \check{q}^d \circ \pi_E$, respectively, commute and span $D|\pi_E^{-1}(V)$. The tangent bundle $\mathscr{T}\Lambda$ is therefore globally spanned by $\xi_{q^1}|\Lambda,\ldots,\xi_{q^d}|\Lambda$. The global parallelism in Λ thus defined is clearly independent of the choice of chart on X/E. Now, let M be a fibre of π_{ED}. The projection onto M of $F|\pi_D^{-1}(M)$ is an involutive complex distribution F_M on M such that $F_M + \overline{F}_M = \mathscr{T}^{\mathbf{C}}M$. Each vector in $\mathscr{T}M$ can be expressed in the form $\mathscr{T}\pi_D(w + \overline{w})$ for some $w \in F$. Let $\mathscr{J}_M\colon \mathscr{T}M \to \mathscr{T}M$ and $h_M\colon \mathscr{T}M \times \mathscr{T}M \to \mathbf{C}$ be defined by

$$\mathscr{J}_M(\mathscr{T}\pi_D(w + \bar{w})) = -\mathscr{T}\pi_D(iw + \overline{iw}) \tag{4.7}$$

and

$$h_M(\mathscr{T}\pi_D(w + \bar{w}), \mathscr{T}\pi_D(w' + \bar{w}')) = 2i\omega(w', \bar{w}). \tag{4.8}$$

Then $\mathscr{J}_M$ is an integrable almost complex structure such that F_M is the distribution of anti-holomorphic vectors, and h_M is a Hermitian form on M with the associated Kähler 2-form given by

$$\omega_M(\mathscr{T}\pi_D(w + \bar{w}), \mathscr{T}\pi_D(w' + \bar{w}')) = \omega(\bar{w}' + w', \bar{w} + w). \tag{4.9}$$

Hence, M has a canonically defined Kähler structure, as required.

A polarization F is said to be *positive* if $i\omega(w,\bar{w}) \geq 0$ for each $w \in F$. For a strongly admissible positive polarization the Kähler metric h_M defined on the fibres M of $\pi_{ED}: X/D \to X/E$ is positive definite. A polarization F is said to be *real* if $F = \bar{F}$. Clearly, a real polarization is positive.

4.2. The bundle $\sqrt{\wedge^n F}$

Given a polarization F of (X,ω), one could take the space of those sections of the line bundle L which are covariantly constant along F to form the representation space. If λ_1 and λ_2 are two such sections of L, their product $\langle\lambda_1,\lambda_2\rangle$ is constant along D and consequently the integral $\int_X \langle\lambda_1,\lambda_2\rangle\, \omega^n$ diverges unless the leaves of D are compact. Since $\langle\lambda_1,\lambda_2\rangle$ defines a function on X/D, one could define a scalar product on X/D if one had a suitable measure. However, X/D has no canonically defined measure. If $\langle\lambda_1,\lambda_2\rangle$ defined a density on the manifold X/D, rather than a scalar function, then one could integrate this density over X/D thereby defining an appropriate scalar product for wave functions. Such a modification can be obtained by using the covariantly constant sections of the tensor product of L with the bundle $\sqrt{\wedge^n F}$ which will be defined presently. We shall see later that the bundle $\sqrt{\wedge^n F}$, apart from providing a suitable scalar product, leads to the correct modifications of the Bohr-Sommerfeld conditions and enables one to obtain unitary representations of groups of canonical transformations generated by certain dynamical variables.

Let F be a polarization of (X,ω). A *linear frame* of F at x is an ordered basis $\underline{w} = (w_1,\ldots,w_n)$ of F_x. The collection of all linear frames of F forms a right principal $GL(n,\mathbb{C})$ fibre bundle $\mathscr{B}F$ over X called the *bundle of linear frames* of F. Associated to $\mathscr{B}F$ is the complex line bundle $\wedge^n F$ over X, the *n'th exterior product* of F. The space of sections μ of $\wedge^n F$ is isomorphic to the space of complex-valued functions $\mu^\#$ on $\mathscr{B}F$ such that

$$\mu^\#(\underline{wC}) = \det(\underline{C}^{-1})\mu^\#(\underline{w}) \tag{4.10}$$

for each $\underline{w} = (w_1,\ldots,w_n) \in \mathscr{B}_x F$ and each $C \in GL(n,\mathbb{C})$; the isomorphism $\mu^\# \mapsto \mu$ being given by

$$\mu(x) = \mu^\#(w_1,\ldots,w_n)w_1\wedge\ldots\wedge w_n. \tag{4.11}$$

Let $ML(n,\mathbb{C})$ denote the double covering group of $GL(n,\mathbb{C})$ and $\rho\colon ML(n,\mathbb{C}) \to GL(n,\mathbb{C})$ the covering homomorphism. $ML(n,\mathbb{C})$ is called the *complex* $n \times n$ *metalinear group*. A *bundle of metalinear frames* of F is a right principal $ML(n,\mathbb{C})$ fibre bundle $\tilde{\mathscr{B}}F$ over X together with a map $\tau\colon \tilde{\mathscr{B}}F \to \mathscr{B}F$ such that the following diagram commutes:

$$\begin{array}{ccc} \tilde{\mathcal{B}}F \times ML(n,\mathbf{C}) & \longrightarrow & \tilde{\mathcal{B}}F \\ \Big\downarrow \tau\times\rho & & \Big\downarrow \tau \\ \mathcal{B}F \times GL(n,\mathbf{C}) & \longrightarrow & \mathcal{B}F \end{array},$$

where the horizontal arrows denote the group actions. The existence of a metalinear frame bundle of F is equivalent to the vanishing of a class in $H^2(X,Z_2)$ characteristic of $\mathcal{B}F$, while the number of distinct metalinear frame bundles is parameterized by the cohomology group $H^1(X,Z_2)$. Let $\chi\colon ML(n,\mathbf{C}) \to \mathbf{C}$ be the unique holomorphic square root of the complex character $\det \circ \rho$ of $ML(n,\mathbf{C})$ such that $\chi(\underline{\tilde{I}}) = 1$, where $\underline{\tilde{I}}$ denotes the identity of $ML(n,\mathbf{C})$. The bundle $\sqrt{\wedge^n F}$ is the fibre bundle over X associated to $\tilde{\mathcal{B}}F$ with standard fibre $\mathbf{C}$ on which $ML(n,\mathbf{C})$ acts by multiplication by $\chi(\underline{\tilde{C}})$, $\underline{\tilde{C}} \in ML(n,\mathbf{C})$. The space of sections ν of $\sqrt{\wedge^n F}$ is isomorphic to the space of complex-valued functions $\nu^\#$ on $\tilde{\mathcal{B}}F$ satisfying the condition

$$\nu^\#(\underline{\tilde{w}}\underline{\tilde{C}}) = \chi(\underline{\tilde{C}}^{-1})\nu^\#(\underline{\tilde{w}}) \tag{4.12}$$

for each $\underline{\tilde{w}} \in \tilde{\mathcal{B}}F$ and each $\underline{\tilde{C}} \in ML(n,\mathbf{C})$. Consequently, by restricting to $\tilde{\mathcal{B}}F_x$, we see that elements $\nu_x \in \sqrt{\wedge^n F_x}$ can be identified with functions $\nu_x^\#\colon \tilde{\mathcal{B}}_x F \to \mathbf{C}$ such that

$$\nu_x^\#(\underline{\tilde{w}}\underline{\tilde{C}}) = \chi(\underline{\tilde{C}}^{-1})\nu_x^\#(\underline{\tilde{w}}) \tag{4.13}$$

for each $\tilde{w} \in \tilde{\mathcal{B}}F_x$ and $\underline{\tilde{C}} \in ML(n,\mathbf{C})$.

A strongly admissible polarization F can be locally spanned by complex Hamiltonian vector fields. Let $\{\xi^1,\ldots,\xi^n\}$ be a set of complex Hamiltonian vector fields spanning F on some open set $W \subseteq X$, and let $\underline{\xi}\colon W \to \mathcal{B}F$ denote the frame field defined by $\underline{\xi}(x) = [\xi^1(x),\ldots,\xi^n(x)]$ for each $x \in X$. Suppose that W is contractible so that there exists a lift of $\underline{\xi}$

to a *metalinear frame field* $\underline{\tilde{\xi}}: W \to \tilde{\mathcal{B}}F$. Let $\nu_{\underline{\tilde{\xi}}}$ be the unique section of $\sqrt{\wedge^n F}$ over W such that

$$\nu_{\underline{\tilde{\xi}}}^{\#} \circ \underline{\tilde{\xi}} = 1. \tag{4.14}$$

Then, every section ν of $\sqrt{\wedge^n F}$ can be represented on W as

$$\nu | W = (\nu^{\#} \circ \underline{\tilde{\xi}}) \nu_{\underline{\tilde{\xi}}}. \tag{4.15}$$

Complex Hamiltonian vector fields in F commute; this can be established by the same argument we used in the previous section to show that Hamiltonian vector fields in D commute. Hence, if $\underline{\xi}$ and $\underline{\xi}'$ are two linear frame fields for F consisting of complex Hamiltonian vector fields defined on some open set W, then $\underline{\xi}'(x) = \underline{\xi}(x)C(x)$ for some function $C: W \to GL(n, \mathbb{C})$ constant along $F|W$. If $\underline{\tilde{\xi}}$ and $\underline{\tilde{\xi}}'$ are lifts of $\underline{\xi}$ and $\underline{\xi}'$, respectively, to metalinear frame fields for $F|W$, then $\underline{\tilde{\xi}}' = \underline{\tilde{\xi}}\tilde{C}$ where $\tilde{C}: W \to ML(n, \mathbb{C})$ is constant along $F|W$. This observation enables us to define an operator ∇ of *partial covariant differentiation* in the direction F acting on the space of sections of $\sqrt{\wedge^n F}$ as follows. A local section ν of $\sqrt{\wedge^n F}$ over W is said to be *covariantly constant* along F if $\nu^{\#} \circ \underline{\tilde{\xi}}$ is constant along F, where $\underline{\tilde{\xi}}$ is any metalinear frame field for F projecting onto a linear frame field consisting of complex Hamiltonian vector fields spanning $F|W$. Since there exist nonvanishing local sections of $\sqrt{\wedge^n F}$ covariantly constant along F, e.g., the section $\nu_{\underline{\tilde{\xi}}}$ defined by Eq. (4.14), and since every section ν of $\sqrt{\wedge^n F}$ can be represented in the form (4.15), we define $\nabla_u \nu$ by

$$(\nabla_u \nu) | W = u(\nu^{\#} \circ \underline{\tilde{\xi}}) \nu_{\underline{\tilde{\xi}}} \tag{4.16}$$

for each $u \in F|W$. This definition does not depend on the choice of $\tilde{\underline{\xi}}$ projecting to a local linear frame field for F consisting of complex Hamiltonian vector fields, and it satisfies all the rules of covariant differentiation if one restricts them to vectors in F. In particular, if ξ and ξ' are vector fields in F, then $[\xi,\xi']$ is in F and, for each section ν of $\sqrt{\wedge^n F}$,

$$(\nabla_\xi \nabla_{\xi'} - \nabla_{\xi'} \nabla_\xi - \nabla_{[\xi,\xi']})\nu = 0. \tag{4.17}$$

4.3. Square integrable wave functions

The connection in L and the operator of partial covariant differentiation of sections of $\sqrt{\wedge^n F}$ in the direction F induce an operator of partial covariant differentiation of sections of $L \otimes \sqrt{\wedge^n F}$ in the direction F. The quantum states of the system under consideration are represented by those sections $\sigma: X \to L \otimes \sqrt{\wedge^n F}$ which are covariantly constant along F. For each complex-valued function ψ on X/D such that the restrictions of ψ to the fibres of $\pi_{ED}: X/D \to X/E$ are holomorphic with respect to the complex structure defined by Eq. (4.7), the section $(\psi \circ \pi_D)\sigma$ of $L \otimes \sqrt{\wedge^n F}$ is also covariantly constant along F. Thus, the quantum states are represented by sections of $L \otimes \sqrt{\wedge^n F}$ which are covariantly constant along D and holomorphic along the fibres of π_{ED}.

Assume that F is a positive polarization, that is, $i\omega(u,\bar{u}) \geq 0$ for each $u \in F$. To each pair (σ_1,σ_2) of sections of $L \otimes \sqrt{\wedge^n F}$ covariantly constant along F we associate a complex density $\langle\sigma_1,\sigma_2\rangle_{X/D}$ on X/D as follows. For each point in X there exists a neighborhood V of this point such that the sections σ_1 and σ_2 restricted to V are of the form

$$\sigma_1|V = \lambda_1 \otimes \nu_1 \quad \text{and} \quad \sigma_2|V = \lambda_2 \otimes \nu_2, \tag{4.18}$$

where λ_1, λ_2 are covariantly constant sections of $L|V$ and ν_1, ν_2 are covariantly constant sections of $\sqrt{\wedge^n F}|V$. Given a point $x \in V$, consider a basis

$$(v_1,\ldots,v_d, u_1,\ldots,u_{n-d},\bar{u}_1,\ldots,\bar{u}_{n-d}, w_1,\ldots,w_d) \tag{4.19}$$

of $\mathscr{T}_x^{\mathbf{C}}X$ such that

$$(v_1,\ldots,v_d) \text{ is a basis of } D_x \tag{4.20}$$

$$\underline{b} = (v_1,\ldots,v_d, u_1,\ldots,u_{n-d}) \text{ is a basis of } F_x \tag{4.21}$$

and, for each $i,j \in \{1,\ldots,d\}$ and each $k,r \in \{1,\ldots,n-d\}$,

$$\omega(v_i,w_j) = \delta_{ij} \tag{4.22}$$

$$i\omega(u_k,\bar{u}_r) = \delta_{kr} \tag{4.23}$$

$$\omega(u_k,w_j) = \omega(w_i,w_j) = 0. \tag{4.24}$$

The basis (4.19) projects to a basis

$$\begin{aligned}(\mathscr{T}\pi_D(u_1),\ldots,\mathscr{T}\pi_D(u_{n-d}),\mathscr{T}\pi_D(\bar{u}_1),\ldots,\mathscr{T}\pi_D(\bar{u}_{n-d}),\\ \mathscr{T}\pi_D(w_1),\ldots,\mathscr{T}\pi_D(w_d))\end{aligned} \tag{4.25}$$

of $\mathscr{T}^{\mathbf{C}}_{\pi_D(x)}X/D$. The value of the density $<\sigma_1,\sigma_2>_{X/D}$ on the basis (4.25) is defined to be

$$<\lambda_1(x),\lambda_2(x)>\nu_1^{\#}(\tilde{\underline{b}})\overline{\nu_2^{\#}(\tilde{\underline{b}})}, \tag{4.26}$$

where $\tilde{\underline{b}} \in \tilde{\mathscr{B}}_x F$ is a metalinear frame of F at x projecting onto the linear frame $\underline{b} \in \mathscr{B}_x F$ given by Eq. (4.21). It can be easily verified that the value of $<\lambda_1(x),\lambda_2(x)>\nu_1^{\#}(\tilde{\underline{b}})\overline{\nu_2^{\#}(\tilde{\underline{b}})}$ depends only on the sections σ_1 and σ_2 and the basis (4.25) of $\mathscr{T}^{\mathbf{C}}_{\pi_D(x)}X/D$. Hence (4.18) - (4.26) define a density on X/D which we denote by $\langle\sigma_1,\sigma_2\rangle_{X/D}$. Note that $\langle\sigma_1,\sigma_2\rangle_{X/D}$ depends linearly on σ_1 and antilinearly on σ_2, and that

$$\langle\sigma,\sigma\rangle_{X/D} \geq 0 \tag{4.27}$$

for each covariantly constant section σ of $L \otimes \sqrt{\wedge^n F}$. Hence, the sesquilinear form

$$(\sigma_1|\sigma_2)_o := \int_{X/D} \langle\sigma_1,\sigma_2\rangle_{X/D} \tag{4.28}$$

defines a Hermitian scalar product on the space of sections of $L \otimes \sqrt{\wedge^n F}$ covariantly constant along F. We denote by $\mathcal{H}^o$ the completion of the pre-Hilbert space of those covariantly constant sections σ for which $(\sigma|\sigma)_o$ is finite. The space $\mathcal{H}^o$ is the subspace of the representation space corresponding to the continuous part of the spectrum of the complete set of commuting observables defining the representation.

If the polarization F is real, $F = \overline{F} = D^{\mathbb{C}}$, and the integral manifolds of D are simply connected, then the space $\mathcal{H}^o$ is equal to the representation space $\mathcal{H}$. In this case one can give the usual local descriptions of sections in $\mathcal{H}^o$ in terms of square integrable complex functions on X/D. Let $(V,\check{q}^1,\ldots,\check{q}^n)$ be a coordinate system on X/D. For each $i \in \{1,\ldots,n\}$, we denote by q^i the pull-back of $\check{q}^i$ to X, that is, $q^i = \check{q}^i \circ \pi_D\colon \pi_D^{-1}(V) \to \mathbb{R}$. The Hamiltonian vector fields $\xi_{q^1},\ldots,\xi_{q^n}$ span $F|\pi_D^{-1}(V)$. We denote by $\underline{\xi}$ the linear frame field of $F|\pi_D^{-1}(V)$ consisting of the Hamiltonian vector fields of $q^1,\ldots,q^n$,

$$\underline{\xi} = (\xi_{q^1},\ldots,\xi_{q^n}), \tag{4.29}$$

and by $\underline{\tilde{\xi}}$ a metalinear frame field projecting onto $\underline{\xi}$.

The section $\nu_{\underline{\tilde{\xi}}}$ of $\sqrt{\wedge^n F}$ defined by

$$\nu_{\underline{\tilde{\xi}}}^{\#} \circ \underline{\tilde{\xi}} = 1 \tag{4.30}$$

is covariantly constant along F.

Let λ_o be a nonvanishing covariantly constant local section of L such that

$$\langle\lambda_o,\lambda_o\rangle = 1. \tag{4.31}$$

Each smooth section $\sigma \in \mathscr{A}^o$ with support in $\pi_D^{-1}(V)$ can be represented in the form

$$\sigma = \psi(q^1,\ldots,q^n)\lambda_o \otimes \nu_{\underline{\tilde{\xi}}}, \tag{4.32}$$

where ψ is a smooth function on R^n with support contained in the range of the coordinate chart $(\check{q}^1,\ldots,\check{q}^n): V \to R^n$. The coordinate functions $\check{q}^1,\ldots,\check{q}^n$ define a density $d^n\check{q}$ on V given by

$$d^n\check{q} = |d\check{q}^1\wedge\ldots\wedge d\check{q}^n|. \tag{4.33}$$

This density coincides with the density $\langle\lambda_o \otimes \nu_{\underline{\tilde{\xi}}}, \lambda_o \otimes \nu_{\underline{\tilde{\xi}}}\rangle_{X/D}$ defined by Eq. (4.26). Hence, the scalar product (4.28) of two sections $\sigma_1 = \psi_1(q^1,\ldots,q^n)\lambda_o \otimes \nu_{\underline{\tilde{\xi}}}$ and $\sigma_2 = \psi_2(q^1,\ldots,q^n)\lambda_o \otimes \nu_{\underline{\tilde{\xi}}}$ with supports in $\pi_D^{-1}(V)$ is given by

$$\begin{aligned}(\psi_1(q^1,\ldots,q^n)\lambda_o \otimes \nu_{\underline{\tilde{\xi}}} \mid \psi_2(q^1,\ldots,q^n)\lambda_o \otimes \nu_{\underline{\tilde{\xi}}}) \\ = \int_V \psi_1(\check{q}^1,\ldots,\check{q}^n)\bar{\psi}_2(\check{q}^1,\ldots,\check{q}^n)\,d^n\check{q}.\end{aligned} \tag{4.34}$$

Thus, the sections $\sigma \in \mathscr{A}^o$ are represented locally by square integrable functions on open sets in X/D.

4.4. Bohr-Sommerfeld conditions

The complement of $\mathscr{H}^o$ in the representation space $\mathscr{H}$ is spanned by generalized (distributional) sections of $L \otimes \sqrt{\wedge^n F}$ which are covariantly constant along F. The supports of these sections are restricted by the Bohr-Sommerfeld conditions as we now explain.

Let Λ be an integral manifold of D. The operator ∇ of partial covariant differentiation of sections of $L \otimes \sqrt{\wedge^n F}$ in the direction F induces a flat connection in the restriction $(L \otimes \sqrt{\wedge^n F})|\Lambda$ of $L \otimes \sqrt{\wedge^n F}$ to Λ. The holonomy group G_Λ of this connection is a subgroup of $\mathbf{C}^\times$. Elements of G_Λ can be obtained by the multiplication of the elements of the holonomy groups of $L|\Lambda$ and $\sqrt{\wedge^n F}|\Lambda$ which correspond to the same loop in Λ. If σ is a section of $L \otimes \sqrt{\wedge^n F}$ covariantly constant along F containing Λ in its domain of definition, then $\sigma|\Lambda$ is a covariantly constant section of $(L \otimes \sqrt{\wedge^n F})|\Lambda$. Parallel transport in $(L \otimes \sqrt{\wedge^n F})|\Lambda$ along loops in Λ results in the multiplication of $\sigma|\Lambda$ by the elements of G_Λ. On the other hand, since $\sigma|\Lambda$ is a covariantly constant section, it does not change under parallel transport. Hence, either $\sigma|\Lambda$ is the zero section, or the holonomy group G_Λ is trivial, i.e., $G_\Lambda = \{1\}$. The union of all integral manifolds Λ of D such that $G_\Lambda = \{1\}$ constitutes the *Bohr-Sommerfeld variety* S. For each $x \in X$, we denote by Λ_x the integral manifold of D passing through x. Then, we have

$$S = \{x \in X \mid G_{\Lambda_x} = \{1\}\}. \tag{4.35}$$

In particular, $S = X$ if each integral manifold Λ of D is simply connected. The preceding discussion implies that the covariantly constant sections of $L \otimes \sqrt{\wedge^n F}$ vanish in the complement of S.

To relate the definition of the variety S to the Bohr-Sommerfeld quantization conditions, consider a contractible open set U in X such that $L|U$ admits a trivializing section λ_o. Then,

$$\nabla\lambda_o = -i\hbar^{-1}\theta \otimes \lambda_o \tag{4.36}$$

for some 1-form θ on U satisfying

$$\omega|U = d\theta \tag{4.37}$$

[cf. Eqs. (3.17) and (3.20)]. For each loop $\gamma\colon [0,1] \to U$, the element of the holonomy group of the connection in L corresponding to γ is given by $\exp(i\hbar^{-1}\int_{\mathrm{Im}\,\gamma}\theta)$. If $\mathrm{Im}\,\gamma$ is contained in an integral manifold Λ of D, we denote by $\exp(-2\pi i d_\gamma)$ the element of the holonomy group of the flat connection in $\sqrt{\wedge^n F}|\Lambda$ corresponding to γ, where the number d_γ is defined up to an arbitrary integer. Thus, for each $\Lambda \subset U$, the condition $G_\Lambda = \{1\}$ is equivalent to

$$\int_{\mathrm{Im}\,\gamma} \theta = (n_\gamma + d_\gamma)h \tag{4.38}$$

for each loop γ in Λ, where n_γ is an integer. Since $\theta = \Sigma p_i dq^i$ for some functions $p_1,\ldots,p_n,q^1,\ldots,q^n$ on U, Eq. (4.38) expresses the *Bohr-Sommerfeld conditions* corrected by the numbers d_γ corresponding to the holonomy group of $\sqrt{\wedge^n F}|\Lambda$.

4.5. Distributional wave functions

A polarization F of (X,ω) is said to be *complete* if the Hamiltonian vector fields contained in D are complete. In this section we analyze the structure of the representation space for a complete strongly admissible real polarization. Simple examples of the general framework presented here are provided by the quantization of a harmonic oscillator in the energy representation and the quantization of charge.

Let F be a complete strongly admissible real polarization of (X,ω). The integral manifolds of D are isomorphic as affine manifolds to the products of tori and affine spaces. For each $k \in \{0,1,\ldots,n\}$, we denote by X_k the subset of X consisting of all integral manifolds of D isomorphic to the product of the k-torus T^k by the affine space R^{n-k}:

$$X_k = \{x \in X \mid \Lambda_x \approx T^k \times R^{n-k}\}, \tag{4.39}$$

and set

$$S_k = S \cap X_k. \tag{4.40}$$

The elements of the structure of the Bohr-Sommerfeld variety S necessary for the definition of a scalar product on the space of the generalized section can be characterized as follows. For each $x \in S_k$, there exists an open neighborhood V of $\pi_D(x)$ in X/D and a k-codimensional submanifold Q of V such that

$$\pi_D(X_k) \cap Q \subseteq \pi_D(S_k) \tag{4.41}$$

and

$$\pi_D(S) \cap V \subseteq Q. \tag{4.42}$$

Also, for $k \geq 1$, there is a canonically defined k-dimensional involutive distribution K on $\pi_D^{-1}(Q)$ contained in

$D|\pi_D^{-1}(Q)$, the integral manifolds of which are diffeomorphic to T^k. The distribution K is endowed with a density κ, invariant under the action of the Hamiltonian vector fields in $D|\pi_D^{-1}(Q)$, which associates to each integral manifold of K the total volume 1.

For each $k \in \{0,1,\dots,n\}$, consider the space of (discontinuous) sections σ of $L \otimes \sqrt{\wedge^n F}$ satisfying the following conditions:

$$\text{support } \sigma \subseteq S_k, \tag{4.43}$$

$$\left.\begin{array}{l}\text{for each } x \in S_k, \text{ the restriction of } \sigma \text{ to} \\ \pi_D^{-1}(Q), \text{ where } Q \text{ satisfies (4.41) and (4.42),} \\ \text{is a smooth section of } (L \otimes \sqrt{\wedge^n F})|\pi_D^{-1}(Q) \\ \text{covariantly constant along } F|\pi_D^{-1}(Q), \text{ and}\end{array}\right\} \tag{4.44}$$

$$\left.\begin{array}{l}\text{the projection to } X/D \text{ of the support of } \sigma \\ \text{is compact.}\end{array}\right\} \tag{4.45}$$

In this space we introduce a scalar product as follows. Let σ_1 and σ_2 be two sections satisfying the conditions (4.43), (4.44) and (4.45). The condition (4.45) implies that there exists a finite number of pairwise disjoint submanifolds $Q_1,\dots,Q_s$ of X/D satisfying (4.41) and (4.42) such that the supports of σ_1 and σ_2 are contained in $\bigcup_{i=1}^{s} \pi_D^{-1}(Q_i)$. The pair (σ_1,σ_2) of sections induces on each Q_i a density $\langle\sigma_1,\sigma_2\rangle_{Q_i}$ which will be defined presently. The scalar product of σ_1 and σ_2, denoted by $(\sigma_1|\sigma_2)_k$, is given by

$$(\sigma_1|\sigma_2)_k = \sum_{i=1}^{s} \int_{Q_i} \langle\sigma_1,\sigma_2\rangle_{Q_i}. \tag{4.46}$$

We denote by $\mathcal{H}^k$ the completion of the space of all sections σ satisfying (4.43) - (4.45) with respect to the scalar product (4.46). For $k = 0$ it coincides with the space $\mathscr{H}^0$ of square integrable sections introduced in Sec. 4.3.

In order to define the density $<\sigma_1,\sigma_2>_{Q_i}$, for $k \geq 1$, consider a point $x \in \pi_D^{-1}(Q_i)$ and a basis $(w_1,\ldots,w_n, v_1,\ldots,v_n)$ of $\mathscr{T}_x X$ such that

$$(w_1,\ldots,w_k) \in K_x \text{ and } \kappa(w_1,\ldots,w_k) = 1 \tag{4.47}$$

$$\underline{w} = (w_1,\ldots,w_n) \in B_x F \tag{4.48}$$

and, for each $i,j \in \{1,\ldots,n\}$,

$$\omega(w_i,v_j) = \delta_{ij}. \tag{4.49}$$

The basis $(w_1,\ldots,w_n, v_1,\ldots,v_n)$ induces a basis $(\mathscr{T}\pi_D(v_{k+1}),\ldots, \mathscr{T}\pi_D(v_n))$ of $\mathscr{T}_{\pi_D(x)}Q_i$. The sections σ_1 and σ_2 can be expressed in a neighborhood of x as tensor products of sections of L and $\sqrt{\wedge^n F}$,

$$\sigma_1 = \lambda_1 \otimes \nu_1 \quad \text{and} \quad \sigma_2 = \lambda_2 \otimes \nu_2. \tag{4.50}$$

The value of the density $<\sigma_1,\sigma_2>_{Q_i}$ on the basis $(\mathscr{T}\pi_D(v_{k+1}),\ldots, \mathscr{T}\pi_D(v_n))$ is given by

$$\begin{aligned} &<\sigma_1,\sigma_2>_{Q_i}(\mathscr{T}\pi_D(v_{k+1}),\ldots, \mathscr{T}\pi_D(v_n)) \\ &\qquad := <\lambda_1(x),\lambda_2(x)>\nu_1^{\#}(\underline{\tilde{w}})\overline{\nu_2^{\#}(\underline{\tilde{w}})}, \end{aligned} \tag{4.51}$$

where $\underline{\tilde{w}} \in \tilde{\mathscr{B}}_x F$ is a metalinear frame for F projecting onto the linear frame $\underline{w}$ given by Eq. (4.48). It can be verified that Eq. (4.51) defines a positive semidefinite density on Q_i which depends linearly on σ_1 and antilinearly on σ_2. Hence, Eq. (4.46) defines a positive definite scalar product.

The full representation space $\mathscr{H}$ is the direct sum of the spaces $\mathscr{H}^k$,

$$\mathcal{H} = \bigoplus_{k=0}^{n} \mathcal{H}^k. \tag{4.52}$$

Elements of $\mathcal{H}$ are sums of sections in the $\mathcal{H}^k$'s, i.e., $\sigma \in H$ is given by $\sigma = \sigma^o + \sigma^1 + \ldots + \sigma^n$, where $\sigma^k \in \mathcal{H}^k$. The scalar product on $\mathcal{H}$ is given by

$$(\sigma_1|\sigma_2) := \sum_{k=0}^{n} (\sigma_1{}^k|\sigma_2{}^k)_k. \tag{4.53}$$

To each connected component of the Bohr-Sommerfeld variety S there corresponds the subspace of $\mathcal{H}$ consisting of sections with supports in this component, and $\mathcal{H}$ can be decomposed into the direct sum of such subspaces. Clearly, $\dim \mathcal{H} = 0$ if S is empty. Conversely, if S is not empty, the representation space is nontrivial, i.e., $\dim \mathcal{H} \geq 1$.

5. BLATTNER-KOSTANT-STERNBERG KERNELS

Let F_1 and F_2 be two polarizations of (X,ω) and $\mathscr{A}_1$ and $\mathscr{A}_2$ the representation spaces corresponding to F_1 and F_2 respectively. For strongly admissible pairs (F_1,F_2) of polarizations there is an intrinsically defined sesqui-linear map $\mathscr{K}_{12}: \mathscr{A}_1 \times \mathscr{A}_2 \to \mathbf{C}$, called the "Blattner-Kostant-Sternberg kernel." The kernel $\mathscr{K}_{12}$ induces a linear map $\mathscr{U}_{12}: \mathscr{A}_2 \to \mathscr{A}_1$ such that

$$\mathscr{K}_{12}(\sigma_1,\sigma_2) = (\sigma_1|\mathscr{U}_{12}\sigma_2) \tag{5.1}$$

for each $\sigma_1 \in \mathscr{A}_1$ and $\sigma_2 \in \mathscr{A}_2$. If $\mathscr{U}_{12}$ is unitary, the representation spaces $\mathscr{A}_1$ and $\mathscr{A}_2$ are said to be "unitarily related."

5.1. Transverse polarizations

Consider first a pair (F_1,F_2) of complete strongly admissible positive polarizations such that

$$F_1 + \overline{F}_2 = \mathscr{T}^{\mathbf{C}}X. \tag{5.2}$$

Each pair (σ_1,σ_2) of sections of $L \otimes \sqrt{\wedge^n F_1}$ and $L \otimes \sqrt{\wedge^n F_2}$, respectively, gives rise to a function $\langle\sigma_1,\sigma_2\rangle$ on X defined

as follows. Given a point $x \in X$, let
$\underline{w}_1 = (w_1^1,\ldots,w_1^n) \in \mathscr{B}_x F_1$ and $\underline{w}_2 = (w_2^1,\ldots,w_2^n) \in \mathscr{B}_x F_2$
be such that, for each $r,s \in \{1,\ldots,n\}$,

$$i\omega(w_1^r,\bar{w}_2^s) = h\delta^{rs} \tag{5.3}$$

where h is Plack's constant; the existence of such a pair of linear frames follows from the transversality condition (5.2). In some neighborhood of x we can factorize σ_1 and σ_2 so that

$$\sigma_1 = \lambda_1 \otimes \nu_1 \quad \text{and} \quad \sigma_2 = \lambda_2 \otimes \nu_2, \tag{5.4}$$

where λ_1 and λ_2 are sections of L, and ν_1 and ν_2 are sections of $\surd\wedge^n F_1$ and $\surd\wedge^n F_2$ respectively. Let, for each $j = 1,2$, $\underline{\tilde{w}}_j \in \tilde{\mathscr{B}}_x F_j$ be a metalinear frame for F_j projecting onto $\underline{w}_j$. We claim that the expression

$$\langle\lambda_1(x),\lambda_2(x)\rangle\nu_1^{\#}(\underline{\tilde{w}}_1)\overline{\nu_2^{\#}(\underline{\tilde{w}}_2)}$$

is defined by $\sigma_1(x)$ and $\sigma_2(x)$ up to the factor ± 1. It is clearly independent of the factorization of σ_1 and σ_2. Moreover, let $\underline{\tilde{w}}_1' \in \tilde{\mathscr{B}}_x F_1$ and $\underline{\tilde{w}}_2' \in \tilde{\mathscr{B}}_x F_2$ be such that the projections $\underline{w}_1'$ of $\underline{\tilde{w}}_1'$ and $\underline{w}_2'$ of $\underline{\tilde{w}}_2'$ satisfy the condition (5.3). Then, there exist $\underline{\tilde{A}}_1, \underline{\tilde{A}}_2 \in ML(n,\mathbf{C})$ such that $\underline{\tilde{w}}_1' = \underline{\tilde{w}}_1\underline{\tilde{A}}_1$ and $\underline{\tilde{w}}_2' = \underline{\tilde{w}}_2\underline{\tilde{A}}_2$. Taking into account Eq. (4.12), we have

$$\nu_1^{\#}(\underline{\tilde{w}}_1')\overline{\nu_2^{\#}(\underline{\tilde{w}}_2')} = \chi(\underline{\tilde{A}}_1^{-1})\overline{\chi(\underline{\tilde{A}}_2^{-1})}\nu_1^{\#}(\underline{\tilde{w}}_1)\overline{\nu_2^{\#}(\underline{\tilde{w}}_2)}.$$

Since χ is the holomorphic square root of $\det \circ \rho\colon ML(n,\mathbf{C}) \to \mathbf{C}$, it follows that $\chi(\underline{\tilde{A}}_j^{-1}) = \pm(\det \underline{A}_j^{-1})^{\frac{1}{2}}$ where $\underline{A}_j = \rho(\underline{\tilde{A}}_j)$. Hence, $\chi(\underline{\tilde{A}}_1^{-1})\overline{\chi(\underline{\tilde{A}}_2^{-1})} = \pm(\det \underline{A}_1)^{-\frac{1}{2}}(\det \underline{\bar{A}}_2)^{-\frac{1}{2}}$. On the other hand, Eq. (5.3) implies that $\underline{A}_1^t\underline{\bar{A}}_2 = \underline{I}$, where $\underline{A}_1^t$ denotes the transpose of $\underline{A}_1$, so that $(\det \underline{A}_1)(\det \underline{\bar{A}}_2) = (\det \underline{A}_1^t)(\det \underline{\bar{A}}_2) = 1$ which proves our assertion.

If there was a way of restricting the arbitrariness in

the choice of the metalinear frames $\underline{\tilde{w}}_1$ and $\underline{\tilde{w}}_2$ such that their projections $\underline{w}_1$ and $\underline{w}_2$, respectively, satisfy Eq. (5.3), one could define a function $\langle\sigma_1,\sigma_2\rangle: X \to \mathbf{C}$ by setting

$$\langle\sigma_1,\sigma_2\rangle(x) = \langle\lambda_1(x),\lambda_2(x)\rangle\nu_1{}^{\#}(\underline{\tilde{w}}_1)\nu_2{}^{\#}(\underline{\tilde{w}}_2). \tag{5.5}$$

Such a restriction is provided by lifting Eq. (5.3) to the metalinear frame bundles. This will be done in terms of the metaplectic structure introduced in Sec. 5.3. In the meantime we assume that $\langle\sigma_1,\sigma_2\rangle$ is defined by Eq. (5.5), where $\underline{\tilde{w}}_1 \in \tilde{\mathscr{B}}_x F_1$ and $\underline{\tilde{w}}_2 \in \tilde{\mathscr{B}}_x F_2$ satisfy condition (5.80) of Sec. 5.3.

Let $\underline{\zeta}_1 = (\zeta_1{}^1,\ldots,\zeta_1{}^n)$ and $\underline{\zeta}_2 = (\zeta_2{}^1,\ldots,\zeta_2{}^n)$ be local linear frame fields for F_1 and F_2, respectively, which consist of (complex) Hamiltonian vector fields, and let $\underline{\tilde{\zeta}}_i$ be a lift of $\underline{\zeta}_i$ to a metalinear frame field for F_i. Assume that the factorizing sections ν_1 and ν_2 in Eq. (5.4) are chosen so that

$$\nu_1{}^{\#} \circ \underline{\tilde{\zeta}}_1 - \nu_2{}^{\#} \circ \underline{\tilde{\zeta}}_2 = 1. \tag{5.6}$$

For each $j,k \in \{1,\ldots,n\}$ set

$$d^{jk} = (i/\hbar)\omega(\zeta_1{}^j,\bar{\zeta}_2{}^k). \tag{5.7}$$

The matrix (d^{jk}) is non-singular and we denote by (d_{km}) the inverse matrix so that

$$\sum_k d^{jk}d_{km} = \delta_m{}^j. \tag{5.8}$$

The frame fields $\underline{\zeta}_1$ for F_1 and

$$\underline{\zeta}_2' := (\sum_k \zeta_2{}^k\overline{d_{k1}},\ldots,\sum_k \zeta_2{}^k\overline{d_{kn}}) \tag{5.9}$$

for F_2 satisfy the condition (5.3) at each point of their common domain. Let $\underline{\tilde{\zeta}}_2{}'$ be a lift of $\underline{\zeta}_2{}'$ to a metalinear frame field for F_2. We have

$$(\nu_1{}^{\#} \circ \underline{\tilde{\zeta}}_1)\overline{(\nu_2{}^{\#} \circ \underline{\tilde{\zeta}}_2{}')} = \det(d_{jk})^{-\frac{1}{2}}, \tag{5.10}$$

where the ambiguity in the sign is absorbed in the choice of the branch of the square root. Taking into account Eqs. (5.6), (5.7) and (5.8), we can rewrite Eq. (5.5) in the form

$$<\sigma_1,\sigma_2> = (i/\hbar)^{n/2}\left[\det \omega(\zeta_1{}^j,\bar{\zeta}_2{}^k)\right]^{\frac{1}{2}}<\lambda_1,\lambda_2>. \tag{5.11}$$

It is apparent from Eq. (5.11) that $<\sigma_1,\sigma_2>$ is a smooth function on X if the sections σ_1 and σ_2 are smooth.

For $\sigma_1 \in \mathscr{A}_1$ and $\sigma_2 \in \mathscr{A}_2$, the kernel $\mathscr{K}_{12}(\sigma_1,\sigma_2)$ is defined by the integration of the function $<\sigma_1,\sigma_2>$. Consider first the case when σ_1 and σ_2 are square integrable sections, i.e., $\sigma_1 \in \mathscr{A}_1^o$ and $\sigma_2 \in \mathscr{A}_2^o$. Then

$$\mathscr{K}_{12}(\sigma_1,\sigma_2) := \int_X <\sigma_1,\sigma_2>|\omega^n|, \tag{5.12}$$

where $|\omega^n|$ is the canonical density on X defined by the n'th exterior power of ω. In order to ensure the convergence of the integral in Eq. (5.12) one needs some additional assumptions on the polarizations F_1 and F_2. We shall not deal with this problem here.

Suppose now that σ_1 and σ_2 are distributional sections with supports in the manifolds $\pi_{D_1}{}^{-1}(Q_1)$ and $\pi_{D_2}{}^{-1}(Q_2)$ respectively, where, for each $i = 1,2$, Q_i satisfies the conditions (4.41) and (4.42) for the polarization F_i. The Blattner-Kostant-Sternberg kernel $\mathscr{K}_{12}(\sigma_1,\sigma_2)$ is defined by the integration of $<\sigma_1,\sigma_2>$ over $\pi_{D_1}{}^{-1}(Q_1) \cap \pi_{D_2}{}^{-1}(Q_2)$ with respect to a density $\delta_{Q_1Q_2}$ (which will be defined presently):

$$\mathscr{K}_{12}(\sigma_1,\sigma_2) := \int \langle\sigma_1,\sigma_2\rangle \delta_{Q_1Q_2}. \tag{5.13}$$

The value of $\mathscr{K}_{12}$ on an arbitrary pair of elements of $\mathscr{A}_1 \times \mathscr{A}_2$ can be obtained from Eqs. (5.12) and (5.13) by linearity.

To define $\delta_{Q_1Q_2}$, let us consider a point $x \in \pi_{D_1}{}^{-1}(Q_1) \cap \pi_{D_2}{}^{-1}(Q_2)$ and a basis of $\mathscr{T}_x[\pi_{D_1}{}^{-1}(Q_1) \cap \pi_{D_2}{}^{-1}(Q_2)]$ of the form

$$(\underline{w}_1,\underline{w}_2,\underline{v}) = (w_1{}^1,\ldots,w_1{}^{k^1},w_2{}^1,\ldots,w_2{}^{k^2},v_1,\ldots,v_m), \tag{5.14}$$

where $k^1 = \operatorname{codim} Q_1$, $k^2 = \operatorname{codim} Q_2$, and $m = 2(n-k^1-k^2)$, such that

$$w_1{}^1,\ldots,w_1{}^{k^1} \in K_1 \text{ and } \kappa_1(w_1{}^1,\ldots,w_1{}^{k^1}) = 1 \tag{5.15}$$

and

$$w_2{}^1,\ldots,w_2{}^{k^2} \in K_2 \text{ and } \kappa_2(w_2{}^1,\ldots,w_2{}^{k^2}) = 1. \tag{5.16}$$

Here, for $i = 1,2$, K_i is the distribution on $\pi_{D_i}{}^{-1}(Q_i)$ such that its integral manifolds are k^i-tori and κ_i is the density on K_i introduced in Sec. 4.5. Let $u_1{}^1,\ldots,u_1{}^{k^1}$, $u_2{}^1,\ldots,u_2{}^{k^2}$ be vectors in $\mathscr{T}_xX$ such that

$$\omega(w_i{}^r,u_k{}^s) = \delta_{ij}\delta^{rs} \text{ and } \omega(u_i{}^r,u_j{}^s) = \omega(u_i{}^r,v_p) = 0 \tag{5.17}$$

for all $i,j \in \{1,2\}$, $r,s \in \{1,\ldots,k^i\}$ and $p \in \{1,\ldots,m\}$. The value of the density $\delta_{Q_1Q_2}$ on the basis $(\underline{w}_1,\underline{w}_2,\underline{v})$ is defined to be the absolute value of ω^n on the basis

$$(\underline{w}_1,\underline{w}_2,\underline{v},\underline{u}_1,\underline{u}_2) = (w_1{}^1,\ldots,w_1{}^{k^1},w_2{}^1,\ldots,w_2{}^{k^2},v_1,\ldots,v_m, u_1{}^1,\ldots,u_1{}^{k^1},u_2{}^1,\ldots,u_2{}^{k^2}) \tag{5.18}$$

of $\mathscr{T}_xX$,

$$\delta_{Q_1Q_2}(\underline{w}_1,\underline{w}_2,\underline{v}) := |\omega^n(\underline{w}_1,\underline{w}_2,\underline{v},\underline{u}_1,\underline{u}_2)|. \tag{5.19}$$

5.2. Strongly admissible pairs of polarizations

Let F_1 and F_2 be complete strongly admissible positive polarizations. Set

$$D_{12} = D_1 \cap D_2 \quad\text{and}\quad E_{12} = E_1 + E_2. \tag{5.20}$$

Then

$$F_1 \cap \overline{F}_2 = D_{12}{}^{\mathbf{C}} \quad\text{and}\quad F_1 + \overline{F}_2 = E_{12}{}^{\mathbf{C}}. \tag{5.21}$$

This can be seen as follows. If $w \in F_1 \cap \overline{F}_2$ then, by the positivity of F_1 and F_2, $i\omega(w,\overline{w}) \geq 0$ and $i\omega(\overline{w},w) \geq 0$. Hence $\omega(\overline{w},w) = 0$ so that $w \in D_1{}^{\mathbf{C}}$ and $w \in D_2{}^{\mathbf{C}}$. Thus $w \in D_{12}{}^{\mathbf{C}}$ and $F_1 \cap \overline{F}_2 \subseteq D_{12}{}^{\mathbf{C}}$. The inclusion $D_{12}{}^{\mathbf{C}} \subseteq F_1 \cap \overline{F}_2$ is obvious. Similarly, the inclusion $F_1 + \overline{F}_2 \subseteq E_{12}{}^{\mathbf{C}}$ is obvious, and the equality follows from the fact that at each point $x \in X$,

$$\dim_{\mathbf{C}}(F_1 + \overline{F}_2) = \operatorname{codim}_{\mathbf{C}}(F_1 \cap \overline{F}_2) = \operatorname{codim}_{\mathbf{C}} D_{12}{}^{\mathbf{C}} = \dim_{\mathbf{C}} E_{12}{}^{\mathbf{C}}.$$

The pair (F_1,F_2) of polarizations is said to be *strongly admissible* if E_{12} is an involutive distribution on X and the spaces X/D_{12} and X/E_{12} of integral manifolds of D_{12} and E_{12}, Respectively, are quotient manifolds of X. We denote by $\pi_{D_{12}} : X \to X/D_{12}$ and $\pi_{E_{12}} : X \to X/E_{12}$ the canonical projections.

In this section we restrict our attention to a strongly admissible pair (F_1,F_2) of polarizations. Let $\mathscr{H}_1$ and $\mathscr{H}_2$ denote the representation spaces associated with the polarizations F_1 and F_2 respectively. Given a pair of sections $\sigma_1 \in \mathscr{H}_1$ and $\sigma_2 \in \mathscr{H}_2$, we will construct a density

$\delta(\sigma_1,\sigma_2)$ on X/D_{12}. The kernel $\mathscr{K}_{12}(\sigma_1,\sigma_2)$ will be given by the integration of this density over X/D_{12},

$$\mathscr{K}_{12}(\sigma_1,\sigma_2) := \int_{X/D_{12}} \delta(\sigma_1,\sigma_2). \tag{5.22}$$

Consider a basis of $\mathscr{T}_x^{\mathbf{C}}X$ of the form $(\underline{v},\underline{u}_1,\underline{u}_2,\underline{\bar{u}}_1,\underline{\bar{u}}_2,\underline{t})$, where

$$\underline{v} \text{ is a basis of } D_{12} \tag{5.23}$$

$$\underline{w}_1 = (\underline{v},\underline{u}_1) \in B_xF_1, \quad \underline{w}_2 = (\underline{v},\underline{u}_2) \in B_xF_2 \tag{5.24}$$

$$i\omega(u_1{}^j,\bar{u}_2{}^k) = \hbar\delta^{jk} \tag{5.25}$$

$$\omega(v^r,t^s) = \delta^{rs} \tag{5.26}$$

$$\omega(u_1{}^j,t^s) = \omega(\bar{u}_1{}^j,t^s) = \omega(u_2{}^j,t^s) = \omega(\bar{u}_2{}^j,t^s) = 0 \tag{5.27}$$

for $j,k \in \{1,\ldots,n-m\}$, $r,s \in \{1,\ldots,m\}$, and $m = \dim D_{12}$. This basis gives rise to a basis

$$\underline{b} = (\mathscr{T}\pi_{12}(\underline{u}_1),\mathscr{T}\pi_{12}(\underline{u}_2),\mathscr{T}\pi_{12}(\underline{\bar{u}}_1),\mathscr{T}\pi_{12}(\underline{\bar{u}}_2),\mathscr{T}\pi_{12}(\underline{t})) \tag{5.28}$$

of $\mathscr{T}_z^{\mathbf{C}}(X/D_{12})$, where $z = \pi_{12}(x)$. We choose the factorization of the sections σ_1 and σ_2 given by Eq. (5.4), and a pair $\underline{\tilde{w}}_1$ and $\underline{\tilde{w}}_2$ of metalinear frames for F_1 and F_2, respectively, such that $\underline{\tilde{w}}_1$ projects onto $\underline{w}_1$ and $\underline{\tilde{w}}_2$ projects onto $\underline{w}_2$. We define the density $\delta(\sigma_1,\sigma_2)$ by the formula

$$\delta(\sigma_1,\sigma_2)(\underline{b}) := \hbar^{n-m}\langle\lambda_1(x),\lambda_2(x)\rangle\nu_1{}^{\#}(\underline{\tilde{w}}_1)\overline{\nu_2{}^{\#}(\underline{\tilde{w}}_2)} \tag{5.29}$$

since, for $m = 0$, $\delta(\sigma_1,\sigma_2)$ coincides with $\langle\sigma_1,\sigma_2\rangle|\omega^n|$.

We now show that Eq. (5.29) does indeed define a density $\delta(\sigma_1,\sigma_2)$ on $\mathscr{T}^{\mathbf{C}}(X/D_{12})$. If $\underline{\tilde{w}}_1{}' = \underline{\tilde{w}}_1\underline{\tilde{A}}_1$ and $\underline{\tilde{w}}_2{}' = \underline{\tilde{w}}_2\underline{\tilde{A}}_2$ are another pair of metalinear frames such that their projections $\underline{w}_1{}' = \underline{w}_1\underline{A}_1$ and $\underline{w}_2{}' = \underline{w}_2\underline{A}_2$ satisfy the conditions (5.23),

(5.24) and (5.25), and $\underline{t}'$ is defined by Eqs. (5.26) and (5.27) with $\underline{v}$, $\underline{u}_1$ and $\underline{u}_2$ replaced by $\underline{v}'$, $\underline{u}_1'$ and $\underline{u}_2'$, we obtain a new basis $\underline{b}'$ of $\mathscr{T}_z^{\mathbf{C}}(X/D_{12})$. To see how $\underline{b}'$ is related to $\underline{b}$, notice that Eqs. (5.23), (5.24) and (5.25) imply that $\underline{A}_1$ and $\underline{A}_2$ are in the block form

$$\underline{A}_j = \begin{bmatrix} \underline{B} & \underline{D}_j \\ 0 & \underline{C}_j \end{bmatrix}, \tag{5.30}$$

where $\underline{B} \in GL(m,\mathbf{R})$ and $\underline{C}_1, \underline{C}_2 \in GL(n-m,\mathbf{C})$, and

$$\underline{C}_1{}^t\overline{\underline{C}}_2 = \underline{I}. \tag{5.31}$$

Thus

$$\underline{b}' = \underline{b}\cdot\underline{M}, \tag{5.32}$$

where

$$\underline{M} = \begin{bmatrix} \underline{C}_1 & 0 & 0 & 0 & 0 \\ 0 & \underline{C}_2 & 0 & 0 & 0 \\ 0 & 0 & \overline{\underline{C}}_1 & 0 & 0 \\ 0 & 0 & 0 & \overline{\underline{C}}_2 & 0 \\ 0 & 0 & 0 & 0 & (\underline{B}^t)^{-1} \end{bmatrix}. \tag{5.33}$$

On the other hand, the same argument that was used in Sec. 5.1 shows that

$$\nu_1{}^{\#}(\underline{\tilde{w}}_1')\overline{\nu_2{}^{\#}(\underline{\tilde{w}}_2')} = \pm(\det \underline{A}_1)^{-\frac{1}{2}}(\det \overline{\underline{A}}_2)^{-\frac{1}{2}}\nu_1{}^{\#}(\underline{\tilde{w}}_1)\nu_2{}^{\#}(\underline{\tilde{w}}_2)$$

$$= \pm(\det \underline{B})^{-1}\nu_1{}^{\#}(\underline{\tilde{w}}_1)\nu_2{}^{\#}(\underline{\tilde{w}}_2) = \pm(\det \underline{M})\nu_1{}^{\#}(\underline{\tilde{w}}_1)\nu_2{}^{\#}(\underline{\tilde{w}}_2).$$

Hence, the right hand side of Eq. (5.29) transforms, up to sign, as a density. The choice of sign is determined by lifting the conditions (5.23), (5.24) and (5.25) to the bundles of metalinear frames, and will be discussed in the following section.

It remains to show that the right hand side of Eq. (5.29) is independent of the choice of x in the fibre $\pi_{12}{}^{-1}(z)$. Let $\underline{\xi} = (\xi^1,\ldots,\xi^m)$ be a linear frame field for D_{12} consisting of Hamiltonian vector fields, and let $\zeta_1{}^1,\ldots,\zeta_1{}^{n-m},\zeta_2{}^1,\ldots,\zeta_2{}^{n-m}$ be (complex) Hamiltonian vector fields such that

$$\underline{\beta}_i = (\xi^1,\ldots,\xi^m,\zeta_i{}^1,\ldots,\zeta_i{}^{n-m}) \tag{5.34}$$

is a linear frame field for F_i in some neighborhood of x. Let $\underline{\tilde{\beta}}_i$ be a lift of $\underline{\beta}_i$ to a metalinear frame field for F_i, and assume that the factorizing sections ν_1 and ν_2 of σ_1 and σ_2, respectively, are chosen so that

$$\nu_1{}^{\#}(\underline{\tilde{\beta}}_1) = \nu_2{}^{\#}(\underline{\tilde{\beta}}_2) = 1. \tag{5.35}$$

Then, ν_1,ν_2 and λ_1,λ_2 are covariantly constant along F_{12}. Let, for each $j,k \in \{1,\ldots,n-m\}$,

$$d^{jk} = (i/\hbar)\omega(\zeta_1{}^j,\bar{\zeta}_2{}^k). \tag{5.36}$$

The matrix (d^{jk}) is non-singular and we denote by (d_{kr}) the inverse matrix so that

$$\sum_{k=1}^{n-m} d^{jk}d_{kr} = \delta_r{}^j. \tag{5.37}$$

The frame fields $\underline{\beta}_1$ for F_1 and

$$\underline{\beta}_2' := (\xi^1,\ldots,\xi^m, \sum_k \zeta_2{}^k\,\bar{d}_{k1},\ldots, \sum_k \zeta_2{}^k\bar{d}_{k,n-m}) \tag{5.38}$$

for F_2 satisfy at each point the conditions (5.23), (5.24) and (5.25), with $u_i{}^j$ replaced by $\zeta_i{}^j$ and v^j replaced by ξ^j, $i = 1,2$, $j = 1,\ldots,n-m$. Let $\underline{\tilde{\beta}}_2'$ be a lift of $\underline{\beta}_2'$ to a metalinear frame field for F_2. We have

$$<\lambda_1,\lambda_2>\nu_1{}^{\#}(\underline{\tilde{\beta}}_1)\overline{\nu_2{}^{\#}(\underline{\tilde{\beta}}_2')} = \pm<\lambda_1,\lambda_2>[\det(d_{jk})]^{-\frac{1}{2}}. \tag{5.39}$$

The frame fields $\underline{\beta}_1$ and $\underline{\beta}_2'$ define a frame field on $\mathcal{F}^{\mathbb{C}}(X/D_{12})$. Since the left hand side of Eq. (5.39) is supposed to be the value of the density $\delta(\sigma_1,\sigma_2)$ on this frame field, the right hand side must be constant along D_{12}. The first factor $\langle\lambda_1,\lambda_2\rangle$ is constant along D_{12} since λ_1 and λ_2 are, by hypothesis, covariantly constant along F_1 and F_2 respectively. Thus, Eq. (5.29) defines a density on X/D_{12} if and only if $\det(d_{jk})$ is constant along D_{12}, which is equivalent to the condition

$$\xi^r\left[\det \omega(\zeta_1{}^j,\bar{\zeta}_2{}^k)\right] = 0 \tag{5.40}$$

for each $r \in \{1,\ldots,m\}$. However, for each $i,j \in \{1,\ldots,n-m\}$ and $r \in \{1,\ldots,m\}$,

$$\xi^r\left[\omega(\zeta_1{}^j,\bar{\zeta}_2{}^k)\right] = 0$$

since ξ^r, $\zeta_1{}^j$ and $\bar{\zeta}_2{}^k$ are Hamiltonian vector fields, and the Hamiltonian vector fields contained in a polarization commute.

We have shown that Eq. (5.29) defines, up to a sign, a density on X/D_{12}. We shall fix the sign by setting some additional restrictions on the choice of $\underline{\tilde{w}}_1$ and $\underline{\tilde{w}}_2$; this will be done in the last paragraph of Sec. 5.4. Making use of Eqs. (5.35), (5.36), (5.37) and (5.39), we can rewrite Eq. (5.29) in the form

$$\begin{aligned}&\delta(\sigma_1,\sigma_2)(\underline{b}) \\ &\quad = (i\hbar)^{(n-m)/2}\left[\det \omega(\zeta_1{}^j(x),\bar{\zeta}_2{}^k(x))\right]^{\frac{1}{2}}\langle\lambda_1(x),\lambda_2(x)\rangle,\end{aligned} \tag{5.41}$$

where the ambiguity in the sign is absorbed in the choice of the branch of the square root.

5.3. Metaplectic structure

A *symplectic frame* at x is an ordered basis $(u_1,\ldots,u_n;v_1,\ldots,v_n)$ of $\mathcal{T}_xX$, denoted in the following by $(\underline{u};\underline{v})$, such that

$$\omega(u_i,v_j) = \delta_{ij},\ \omega(u_i,u_j) = \omega(v_i,v_j) = 0 \qquad (5.42)$$

for $i,j \in \{1,\ldots,n\}$. The collection of all symplectic frames at all points of X forms a right principal $Sp(n,\mathbf{R})$ bundle $\mathcal{B}_\omega\mathcal{T}X$ over X, referred to as the *symplectic frame bundle* of (X,ω), where $Sp(n,\mathbf{R})$ denotes the $n \times n$ symplectic group. The group $Sp(n,\mathbf{R})$ can be realized explicitly as the subgroup of $GL(2n,\mathbf{R})$ consisting of matrices of the block form

$$\begin{bmatrix} \underline{T}_1 & \underline{T}_2 \\ \underline{T}_3 & \underline{T}_4 \end{bmatrix}, \qquad (5.43)$$

where $\underline{T}_1$, $\underline{T}_2$, $\underline{T}_3$, $\underline{T}_4$ are $n \times n$ matrices satisfying

$$\underline{T}_4^t\underline{T}_1 - \underline{T}_2^t\underline{T}_3 = \underline{I},\quad \underline{T}_1^t\underline{T}_3 = \underline{T}_3^t\underline{T}_1,\quad \underline{T}_2^t\underline{T}_4 = \underline{T}_4^t\underline{T}_2. \qquad (5.44)$$

Here, the superscript t denotes transposition. $Sp(n,\mathbf{R})$ is homeomorphic to the Cartesian product of the n-dimensional unitary group and a Euclidean space. Hence, its fundamental group is isomorphic to the additive group $\mathbf{Z}$ of integers.

Let $Mp(n,\mathbf{R})$ denote the double covering group of $Sp(n,\mathbf{R})$ and $\rho\colon Mp(n,\mathbf{R}) \to Sp(n,\mathbf{R})$ the covering homomorphism. The group $Mp(n,\mathbf{R})$ is called the $n \times n$ *metaplectic group.* A *metaplectic frame bundle* of (X,ω) is a right principal $Mp(n,\mathbf{R})$ bundle $\tilde{\mathcal{B}}_\omega\mathcal{T}X$ over X together with a map

$\tau: \tilde{\mathscr{B}}_\omega \mathscr{T}X \to \mathscr{B}_\omega \mathscr{T}X$ such that the following diagram commutes:

$$\begin{array}{ccc} \tilde{\mathscr{B}}_\omega \mathscr{T}X \times Mp(n,\mathbf{R}) & \longrightarrow & \tilde{\mathscr{B}}_\omega \mathscr{T}X \\ \Big\downarrow \tau \times \rho & & \Big\downarrow \tau \\ \mathscr{B}_\omega \mathscr{T}X \times Sp(n,\mathbf{R}) & \longrightarrow & \mathscr{B}_\omega \mathscr{T}X \end{array} . \tag{5.45}$$

The horizontal arrows denote the group actions. The existence of a metaplectic frame bundle of (X,ω) is equivalent to the vanishing of a class in $H^2(X,Z_2)$ characteristic of $\mathscr{B}_\omega \mathscr{T}X$, and the set of all $Mp(n,\mathbf{R})$-equivariant equivalence classes of metaplectic frame bundles, provided it is not empty, can be parametrized by $H^1(X,Z_2)$.

The purpose of introducing metaplectic frame bundles is to allow as to assign metalinear structures to all positive polarizations simultaneously. In other words, the choice of a metaplectic frame bundle uniquely determines a metalinear frame bundle for each positive polarization. This is accomplished via the intermediary of the "bundle of positive Lagrangian frames", which we now define.

A *positive Lagrangian frame* at $x \in X$ is an ordered n-tuple $\underline{w} = (w_1,\ldots,w_n)$ of vectors in $\mathscr{T}_x^{\mathbf{C}}X$ such that $w_1,\ldots,w_n$ are linearly independent over $\mathbf{C}$ and

$$\omega(w_j,w_k) = 0 \quad \text{and} \quad i\omega(w_k,\overline{w}_k) \geq 0 \tag{5.46}$$

for all $j,k = \{1,\ldots,n\}$. If F is a positive polarization, then each element of $\mathscr{B}F$ is a positive Lagrangian frame of (X,ω). The collection of all positive Lagrangian frames at all points of X forms the *bundle of positive Lagrangian frames* $\mathscr{P}_\omega \mathscr{T}X$ of (X,ω). We now describe the structure of $\mathscr{P}_\omega \mathscr{T}X$. Given a symplectic frame $(\underline{u};\underline{v})$ at x, each positive Lagrangian frame $\underline{w}$ can be uniquely expressed as

$$\underline{w} = \underline{u}\,\underline{U} + \underline{v}\,\underline{V} = (\underline{u};\underline{v})\begin{bmatrix} \underline{U} \\ \underline{V} \end{bmatrix}, \tag{5.47}$$

where $\underline{U}$ and $\underline{V}$ are complex $n \times n$ matrices such that

$$\operatorname{rank}\begin{bmatrix} \underline{U} \\ \underline{V} \end{bmatrix} = n, \quad \underline{U}^t\underline{V} = \underline{V}^t\underline{U} \tag{5.48}$$

and

$$i(\underline{V}^\dagger\underline{U} - \underline{U}^\dagger\underline{V}) \text{ is positive semidefinite,} \tag{5.49}$$

where $\underline{U}^\dagger = \overline{\underline{U}}^t$.

Let P denote the set of complex $2n \times n$ matrices $\begin{bmatrix} \underline{U} \\ \underline{V} \end{bmatrix}$ such that (5.48) and (5.49) are satisfied. The symplectic group $Sp(n,\mathbf{R})$ acts on P on the left by matrix multiplication

$$\begin{bmatrix} \underline{T}_1 & \underline{T}_2 \\ \underline{T}_3 & \underline{T}_4 \end{bmatrix}\begin{bmatrix} \underline{U} \\ \underline{V} \end{bmatrix} = \begin{bmatrix} \underline{T}_1\underline{U} + \underline{T}_2\underline{V} \\ \underline{T}_3\underline{U} + \underline{T}_4\underline{V} \end{bmatrix}. \tag{5.50}$$

Eq. (5.47) defines a bijection between $\mathscr{P}_\omega\mathscr{T}_x X$ and P. Moreover, if

$$(\underline{u}';\underline{v}') = (\underline{u};\underline{v})\begin{bmatrix} \underline{T}_1 & \underline{T}_2 \\ \underline{T}_3 & \underline{T}_4 \end{bmatrix} \tag{5.51}$$

is another symplectic frame at x, then

$$\underline{w} = \underline{u}'\underline{U}' + \underline{v}'\underline{V}' = (\underline{u}';\underline{v}')\begin{bmatrix} \underline{U}' \\ \underline{V}' \end{bmatrix}, \tag{5.52}$$

where

$$\begin{bmatrix} \underline{U}' \\ \underline{V}' \end{bmatrix} = \begin{bmatrix} \underline{T}_1 & \underline{T}_2 \\ \underline{T}_3 & \underline{T}_4 \end{bmatrix}^{-1}\begin{bmatrix} \underline{U} \\ \underline{V} \end{bmatrix}. \tag{5.53}$$

Hence, the bundle of Lagrangian frames $\mathscr{P}_\omega\mathscr{T}X$ is a fibre bundle over X with standard fibre P and structure group $Sp(n,\mathbf{R})$, acting on P according to (5.50), which is associated to the bundle of symplectic frames $\mathscr{B}_\omega\mathscr{T}X$. Eq. (5.47) also establishes a bijection between the space of positive Lagrangian frames $\underline{w}$ at $x \in X$ and the space of functions $\underline{w}^\# : \mathscr{B}_\omega\mathscr{T}_x X \to P$ such that, for each $(\underline{u};\underline{v}) \in \mathscr{B}_\omega\mathscr{T}_x X$ and each $g \in Sp(n,\mathbf{R})$

$$\underline{w}^{\#}((\underline{u};\underline{v})g) = g^{-1}\underline{w}^{\#}(\underline{u};\underline{v}), \tag{5.54}$$

given by

$$\underline{w} = (\underline{u};\underline{v})\ \underline{w}^{\#}(\underline{u};\underline{v}). \tag{5.55}$$

In the following we shall identify positive Lagrangian frames with the corresponding functions on the fibres of the bundle of symplectic frames.

Since elements of $\mathcal{P}_{\omega}\mathcal{T}X$ are complex linear n-frames, there is a right action of $GL(n,\mathbb{C})$ on $\mathcal{P}_{\omega}\mathcal{T}X$ such that, for each $\underline{w} = (w_1,\ldots,w_n) \in \mathcal{P}_{\omega}\mathcal{T}X$ and each $\underline{C} = (c_{ij}) \in GL(n,\mathbb{C})$, $\underline{w}\ \underline{C} = (\sum_{i=1}^{n} w_i c_{i1},\ldots,\sum_{i=1}^{n} w_i c_{in})$. The action of $GL(n,\mathbb{C})$ on P corresponding to this action on $\mathcal{P}_{\omega}\mathcal{T}X$ is given by $\left(\begin{bmatrix}\underline{U}\\ \underline{V}\end{bmatrix},\underline{C}\right) \to \begin{bmatrix}\underline{U}\,\underline{C}\\ \underline{V}\,\underline{C}\end{bmatrix}$. For each $\begin{bmatrix}\underline{U}\\ \underline{V}\end{bmatrix} \in P$, the relations (5.48) and (5.49) imply that the matrix $\underline{C}$ defined by

$$\underline{C} = \underline{U} - i\underline{V} \tag{5.56}$$

is non-singular, i.e., $\underline{C} \in GL(n,\mathbb{C})$, and that the matrix $\underline{W}$ defined by

$$\underline{W} = (\underline{U} + i\underline{V})(\underline{U} - i\underline{V})^{-1} \tag{5.57}$$

is symmetric. Moreover, it follows from (5.49) that

$$\|\underline{W}\| \leq 1, \tag{5.58}$$

where $\|\underline{W}\|$ denotes the operator norm of $\underline{W}$. Let B denote the closed unit ball in the space of complex symmetric $n \times n$ matrices, i.e.,

$$B = \{\underline{W} \in \mathbb{C}_n^n \mid \underline{W}^t = W,\ \|\underline{W}\| \leq 1\}. \tag{5.59}$$

The mapping $\begin{bmatrix}\underline{U}\\ \underline{V}\end{bmatrix} \to (\underline{W},\underline{C})$ defined by (5.56) and (5.57) is a bijection from P onto $B \times GL(n,\mathbb{C})$ with inverse given by

$$\underline{U} = \frac{1}{2}(\underline{I} + \underline{W})\underline{C}, \qquad \underline{V} = \frac{i}{2}(\underline{I} - \underline{W})\underline{C}. \tag{5.60}$$

We shall use this bijection to identify P with $B \times GL(n,\mathbf{C})$. The action of $GL(n,\mathbf{C})$ on $B \times GL(n,\mathbf{C})$ corresponding to the action of $GL(n,\mathbf{C})$ on P is given by

$$((\underline{W},\underline{C})\underline{C}') \mapsto (\underline{W},\underline{CC}'). \tag{5.61}$$

Thus, the bijection $P \approx B \times GL(n,\mathbf{C})$ exhibits in P the structure of a $GL(n,\mathbf{C})$-principal fibre bundle over B. The left action of $Sp(n,\mathbf{R})$ on P transferred to $B \times GL(n,\mathbf{C})$ takes on the following form, for each $g = \begin{bmatrix} \underline{T}_1 & \underline{T}_2 \\ \underline{T}_3 & \underline{T}_4 \end{bmatrix}$ in $Sp(n,\mathbf{R})$ and each $(\underline{W},\underline{C}) \in B \times GL(n,\mathbf{C})$:

$$g(\underline{W},\underline{C}) = (g\underline{W},\alpha(g,\underline{W})\underline{C}) \tag{5.62}$$

where

$$\begin{aligned} g\underline{W} = {} & [(\underline{T}_1+i\underline{T}_3)(\underline{I}+\underline{W}) - (\underline{T}_4-i\underline{T}_2)(\underline{I}-\underline{W})] \\ & \times [(\underline{T}_1-i\underline{T}_3)(\underline{I}+\underline{W}) + (\underline{T}_4+i\underline{T}_2)(\underline{I}-\underline{W})]^{-1} \end{aligned} \tag{5.63}$$

and

$$\alpha(g,\underline{W}) = \frac{1}{2}[(\underline{T}_1-i\underline{T}_3)(I+\underline{W}) + (\underline{T}_4+i\underline{T}_2)(\underline{I}-\underline{W})] \in GL(n,\mathbf{C}). \tag{5.64}$$

The mapping $\alpha\colon Sp(n,\mathbf{R}) \times B \to GL(n,\mathbf{C})$ defined by Eq. (5.64) satisfies

$$\alpha(g_1g_2,\underline{W}) = \alpha(g_1,g_2\underline{W})\alpha(g_2,\underline{W}). \tag{5.65}$$

The group $U(n)$ of $n \times n$ unitary matrices can be imbedded in $Sp(n,\mathbf{R})$ as follows:

$$U(n) \ni \underline{S} + i\underline{T} \to \begin{bmatrix} \underline{S} & \underline{T} \\ -\underline{T} & \underline{S} \end{bmatrix} \in Sp(n,\mathbf{R}), \tag{5.66}$$

where $\underline{S}$ and $\underline{T}$ are real matrices satisfying $\underline{S}^t\underline{S} + \underline{T}^t\underline{T} = \underline{I}$, and $\underline{S}^t\underline{T} = \underline{T}^t\underline{S}$. If $g \in Sp(n,\mathbf{R})$ is the image of a unitary matrix $\underline{S} + i\underline{T}$ under the imbedding (5.66), then

$$\alpha(g,\underline{W}) = \underline{S} + i\underline{T}. \tag{5.67}$$

There exists a unique lift $\tilde{\alpha}$: $Mp(n,\mathbf{R}) \times B \to ML(n,\mathbf{C})$ of α such that the diagram

$$\begin{array}{ccc} Mp(n,\mathbf{R}) \times B & \xrightarrow{\tilde{\alpha}} & ML(n,\mathbf{C}) \\ {\scriptstyle \rho \times id} \downarrow & & \downarrow {\scriptstyle \rho} \\ Sp(n,\mathbf{R}) \times B & \xrightarrow{\alpha} & GL(n,\mathbf{C}) \end{array} \tag{5.68}$$

commutes and, for each $\underline{W} \in B$, $\tilde{\alpha}(\tilde{e},\underline{W})$ is the identity in $ML(n,\mathbf{C})$, where $\tilde{e}$ denotes the identity in $Mp(n,\mathbf{R})$. Let $\tilde{P} = B \times ML(n,\mathbf{C})$. Then $\tilde{P}$ is a trivial principal $ML(n,\mathbf{C})$ bundle over B, and there is a left action of $Mp(n,\mathbf{R})$ on $\tilde{P}$ defined by $\tilde{\alpha}$ as follows:

$$\tilde{g}(\underline{W},\underline{\tilde{C}}) = (\rho(\tilde{g})\underline{W},\ \tilde{\alpha}(\tilde{g},\underline{W})\underline{\tilde{C}}) \tag{5.69}$$

for each $\tilde{g} \in Mp(n,\mathbf{R})$ and each $(\underline{W},\underline{\tilde{C}}) \in \tilde{P}$. $\tilde{P}$ is a double covering space of P with covering map τ: $\tilde{P} \to P$ given by $\tau(\underline{W},\underline{\tilde{C}}) = \begin{bmatrix} \underline{U} \\ \underline{V} \end{bmatrix}$, where $\underline{U} = \frac{1}{2}(\underline{I} + \underline{W})\rho(\underline{\tilde{C}})$ and $\underline{V} = \frac{i}{2}(\underline{I} - \underline{W})\rho(\underline{\tilde{C}})$.

The *bundle of metalinear positive Lagrangian frames* $\tilde{\mathscr{P}}_\omega\mathscr{T}X$ is the fibre bundle over X with typical fibre $\tilde{P}$, on which $Mp(n,\mathbf{R})$ acts by (5.69), associated to the metaplectic frame bundle $\tilde{\mathscr{B}}_\omega\mathscr{T}X$. That is, a metalinear Lagrangian frame $\underline{\tilde{w}} \in \tilde{\mathscr{P}}_\omega\mathscr{T}_xX$ can be identified with a function $\underline{\tilde{w}}^\#: \tilde{\mathscr{B}}_\omega\mathscr{T}_xX \to \tilde{P}$ such that, for each metaplectic frame $(\underline{u};\underline{v}) \in \tilde{\mathscr{B}}_\omega\mathscr{T}_xX$ and each $\tilde{g} \in Mp(n,\mathbf{R})$,

$$\underline{\tilde{w}}^\#((\underline{u};\underline{v})\tilde{g}) = \tilde{g}^{-1}\underline{w}^\#(\underline{u};\underline{v}). \tag{5.70}$$

The bundle $\tilde{\mathscr{P}}_\omega\mathscr{T}X$ is a double covering of the bundle $\mathscr{P}_\omega\mathscr{T}X$ of positive Lagrangian frames with covering map $\tau: \tilde{\mathscr{P}}_\omega \mathscr{T}X \to \mathscr{P}_\omega \mathscr{T}X$ given as follows: for each $\tilde{w} \in \tilde{\mathscr{P}}_\omega \mathscr{T}_x X$, $\underline{w} = \tau(\tilde{\underline{w}})$ is the unique element of $\mathscr{P}_\omega \mathscr{T}_x X$ such that the following diagram commutes

$$\begin{array}{ccc} \tilde{\mathscr{B}}_\omega \mathscr{T}_x X & \xrightarrow{\tilde{\underline{w}}^\#} & \tilde{P} \\ \tau \big\downarrow & & \big\downarrow \tau \\ \mathscr{B}_\omega \mathscr{T}_x X & \xrightarrow{\underline{w}^\#} & P \end{array} \quad . \tag{5.71}$$

The right action of $GL(n,\mathbf{C})$ and the left action of $Sp(n,\mathbf{R})$ on P commute with each other. Similarly, the right action of $ML(n,\mathbf{C})$ on $\tilde{P}$ commutes with the left action of $Mp(n,\mathbf{R})$ on $\tilde{P}$. Therefore, $\tilde{\mathscr{P}}_\omega \mathscr{T}X$ inherits a right action of $ML(n,\mathbf{C})$ and the diagram

$$\begin{array}{ccc} \tilde{\mathscr{P}}_\omega \mathscr{T}X \times ML(n,\mathbf{C}) & \longrightarrow & \tilde{\mathscr{P}}_\omega \mathscr{T}X \\ \big\downarrow \tau \times \rho & & \big\downarrow \tau \\ \mathscr{P}_\omega \mathscr{T}X \times GL(n,\mathbf{C}) & \longrightarrow & \mathscr{P}_\omega \mathscr{T}X \end{array} \quad , \tag{5.72}$$

in which the horizontal arrows denote the group actions, commutes.

Let F be a positive polarization of (X,ω). The bundle $\mathscr{B}F$ of linear frames of F is a subbundle of $\mathscr{P}_\omega \mathscr{T}X$ invariant under the action of $GL(n,\mathbf{C})$. The inverse image of $\mathscr{B}F$ under the double covering $\tau: \tilde{\mathscr{P}}_\omega \mathscr{T}X \to \mathscr{P}_\omega \mathscr{T}X$ is a subbundle $\tilde{\mathscr{B}}F$ of $\tilde{\mathscr{P}}_\omega \mathscr{T}X$ invariant under the action of $ML(n,\mathbf{C})$, and τ restricted to $\tilde{\mathscr{B}}F$ defines a double covering denoted by $\tau: \tilde{\mathscr{B}}F \to \mathscr{B}F$. It follows that $\tilde{\mathscr{B}}F$ principal $ML(n,\mathbf{C})$ fibre bundle and that the following diagram commutes

$$\begin{array}{ccc} \tilde{\mathscr{B}}F \times ML(n,\mathbf{C}) & \longrightarrow & \tilde{\mathscr{B}}F \\ \downarrow{\scriptstyle \tau \times \rho} & & \downarrow{\scriptstyle \tau} \\ \mathscr{B}F \times GL(n,\mathbf{C}) & \longrightarrow & \mathscr{B}F \end{array}, \tag{5.73}$$

where the horizontal arrows denote the group actions. The bundle $\tilde{\mathscr{B}}F$ is the metalinear frame bundle of F *induced* by the metaplectic frame bundle $\tilde{\mathscr{B}}_\omega \mathscr{T}X$. In the following we assume that we have chosen a metaplectic frame bundle for (X,ω) and, for each positive polarization F, we consider only the induced metalinear frame bundle.

We can extend the condition (5.3) to arbitrary positive Lagrangian frames $\underline{w}_1 = (w_1{}^1,\ldots,w_1{}^n)$ and $\underline{w}_2 = (w_2{}^1,\ldots,w_2{}^n)$, obtaining a relation in $\mathscr{P}_\omega \mathscr{T}X$ given by

$$i\omega(w_1{}^j, \bar{w}_2{}^k) = h\delta_{jk} \tag{5.74}$$

for all $j,k \in \{1,\ldots,n\}$. Let $(\underline{u};\underline{v}) \in \mathscr{B}_\omega \mathscr{T}X$ and $\begin{bmatrix} \underline{U}_i \\ \underline{V}_i \end{bmatrix} \in P$ be such that $\underline{w}_i = \underline{u}\,\underline{U}_i + \underline{v}\,\underline{V}_i$, for $i = 1,2$. Then Eq. (5.74) reads

$$\underline{V}_2{}^\dagger \underline{U}_1 - \underline{U}_2{}^\dagger \underline{V}_1 = -ih\,\underline{I}. \tag{5.75}$$

If $(\underline{W}_1,\underline{C}_1)$ and $(\underline{W}_2,\underline{C}_2)$ are given by $\underline{W}_j = (\underline{U}_j + i\underline{V}_j)(\underline{U}_j - i\underline{V}_j)^{-1}$ and $\underline{C}_j = \underline{U}_j - i\underline{V}_j$, $j = 1,2$, the condition (5.75) transferred to $B \times GL(n,\mathbf{C})$ takes the form

$$\underline{I} - \underline{W}_2{}^\dagger \underline{W}_1 = 2h(\underline{C}_2{}^\dagger)^{-1}\underline{C}_1{}^{-1}. \tag{5.76}$$

Let B_o be the set of all $n \times n$ complex matrices $\underline{S}$ such that $\|\underline{S}\| \leq 1$ and such that 1 is not an eigenvalue of $\underline{S}$, and consider the map

$$\gamma: B_o \to GL(n,\mathbf{C}): \quad \underline{S} \mapsto \underline{I} - \underline{S}. \tag{5.77}$$

Since, for any $n \times n$ complex matrix $\underline{S}$, $\|\underline{S}\| < 1$ implies that 1 is not an eigenvalue of $\underline{S}$, B_o is contractible and consequently there exists a unique map $\tilde{\gamma}: B_o \to ML(n,\mathbf{C})$ such that

$$\rho \circ \tilde{\gamma} = \gamma \quad \text{and} \quad \tilde{\gamma}(\underline{0}) = \tilde{\underline{I}}. \tag{5.78}$$

We are now in a position to lift the condition (5.76), and hence the condition (5.3), to the bundle of metalinear Lagrangian frames. Let $\tilde{\underline{w}}_1$ and $\tilde{\underline{w}}_2$ be two positive metalinear Lagrangian frames, $(\underline{u};\underline{v})$ a metaplectic frame and, for each $i = 1,2$,

$$(\underline{W}_i, \tilde{\underline{C}}_i) = \tilde{\underline{w}}_i^{\#}(\underline{u};\underline{v}). \tag{5.79}$$

The condition (5.76) is lifted to the bundle $\tilde{\mathscr{P}}_\omega \mathscr{T}X$ by requiring that

$$\tilde{\gamma}(\underline{W}_2^{\dagger}\underline{W}_1) = 2\hbar(\tilde{\underline{C}}_2^{\dagger})^{-1}\tilde{\underline{C}}_1^{-1}. \tag{5.80}$$

Note that operations of Hermitian conjugation and multiplication by a positive real number in $GL(n,\mathbf{C})$ lift to $ML(n,\mathbf{C})$ so that the right hand side of Eq. (5.80) makes sense. Moreover, if $\tilde{g}$ is an element of the kernel of $\rho: Mp(n,\mathbf{R}) \to Sp(n,\mathbf{R})$ so that $e = \rho(\tilde{g})$ is the identity in $Sp(n,\mathbf{R})$, we have $\tilde{\alpha}(\tilde{g},\underline{W}) = \tilde{g}$ for each $\underline{W} \in B$ and

$$\begin{aligned}
2((\tilde{\alpha}(\tilde{g},\underline{W}_2)\tilde{\underline{C}}_2)^{\dagger})^{-1}(\tilde{\alpha}(\tilde{g},\underline{W}_1)\tilde{\underline{C}}_1)^{-1} &= 2((\tilde{g}\tilde{\underline{C}}_2)^{\dagger})^{-1}(\tilde{g}\tilde{\underline{C}}_1)^{-1} \\
= 2(\tilde{g}^{\dagger})^{-1}(\tilde{\underline{C}}_2^{\dagger})^{-1}\tilde{\underline{C}}_1^{-1}\tilde{g}^{-1} &= 2\tilde{g}^{-1}(\tilde{\underline{C}}_2^{\dagger})^{-1}\tilde{\underline{C}}_1^{-1}\tilde{g}^{-1} \\
&= 2(\tilde{\underline{C}}_2^{\dagger})^{-1}\tilde{\underline{C}}_1^{-1}.
\end{aligned}$$

Therefore, Eq. (5.80) is invariant under the action of the kernel of $\rho: Mp(n,\mathbf{R}) \to Sp(n,\mathbf{R})$. Furthermore, Eq. (5.80) projected to $GL(n,\mathbf{C})$ yields Eq. (5.76) which is invariant under the action of $Sp(n,\mathbf{R})$. Hence, Eq. (5.80) is invariant under the action of $Mp(n,\mathbf{R})$ so that it defines a relation in $\tilde{\mathcal{P}}_\omega \mathcal{T}X$ independent of the metaplectic frame used in Eq. (5.79).

5.4. Induced metaplectic structure

Let (F_1,F_2) be a strongly admissible pair of complete strongly admissible positive polarizations such that the distribution D_{12} is orientable. For each integral manifold M of E_{12}, its projection $N = \pi_{D_{12}}(M)$ to X/D_{12} is a submanifold of X/D_{12} endowed with a symplectic form ω_N such that the restriction of ω to M is equal to the pull-back of ω_N to M by the map $\pi_{D_{12}}|M\colon M \to N$,

$$\omega|M = (\pi_{D_{12}}|M)^* \omega_N. \tag{5.81}$$

The metaplectic structure on (X,ω) induces a metaplectic structure on (N,ω_N) which will be used to lift the conditions (5.23), (5.24) and (5.25) to the bundle of metalinear Lagrangian frames.

Given a point $x \in X$, let $\mathcal{B}_\omega^{12}\mathcal{T}_x X$ be the collection of symplectic frames in $\mathcal{B}_\omega \mathcal{T}_x X$ of the form

$$(\underline{s},\underline{u};\underline{t},\underline{v}) = (s^1,\ldots,s^m,u^1,\ldots,u^{n-m};\; t^1,\ldots,t^m,\, v^1,\ldots,v^{n-m}) \tag{5.82}$$

such that

$$s^1,\ldots,s^m \in D_{12} \tag{5.83}$$

and

$$s^1,\ldots,s^m,u^1,\ldots,u^{n-m},\, v^1,\ldots,v^{n-m} \in E_{12}. \tag{5.84}$$

The collection of all the symplectic frames of the form (5.82) which satisfy (5.83) and (5.84) is a subbundle

$$\mathscr{B}_\omega{}^{12}\mathscr{T}X = \bigcup_{x\in X} \mathscr{B}_\omega{}^{12}\mathscr{T}_x X \tag{5.85}$$

of the symplectic frame bundle $\mathscr{B}_\omega \mathscr{T}X$. It is a principal fibre bundle with structure group G consisting of all matrices in $Sp(n,\mathbf{R})$ of the form

$$\begin{bmatrix} \underline{A}^t & \underline{B} & \underline{C} & \underline{D} \\ \underline{0} & \underline{T}_1 & \underline{E} & \underline{T}_2 \\ \underline{0} & \underline{0} & \underline{A}^{-1} & \underline{0} \\ \underline{0} & \underline{T}_3 & \underline{F} & \underline{T}_4 \end{bmatrix}, \tag{5.86}$$

where $\underline{A} \in GL(n,\mathbf{R})$ and $\underline{T}_1, \underline{T}_2, \underline{T}_3, \underline{T}_4$ are $(n-m) \times (n-m)$ matrices satisfying Eq. (5.44). We denote by $\tilde{\mathscr{B}}_\omega{}^{12}\mathscr{T}X$ the inverse image of $\mathscr{B}_\omega{}^{12}\mathscr{T}X$ under the covering map $\tau\colon \tilde{\mathscr{B}}_\omega\mathscr{T}X \to \mathscr{B}_\omega\mathscr{T}X$ and by $\tilde{G}$ the inverse image of G under $\rho\colon Mp(n,\mathbf{R}) \to Sp(n,\mathbf{R})$. For $n > m$, $\tilde{G}$ is a connected subgroup of $Mp(n,\mathbf{R})$. The bundle $\mathscr{B}_\omega{}^{12}\mathscr{T}X$ is a right $\tilde{G}$-principal fibre bundle.

Let G_o be the subgroup of G consisting of matrices of the form

$$\begin{bmatrix} \underline{A}^t & \underline{B} & \underline{C} & \underline{D} \\ \underline{0} & \underline{I} & \underline{E} & \underline{0} \\ \underline{0} & \underline{0} & \underline{A}^{-1} & \underline{0} \\ \underline{0} & \underline{0} & \underline{F} & \underline{I} \end{bmatrix}. \tag{5.87}$$

It is a normal subgroup of G and the quotient G/G_o is isomorphic to $Sp(n-m,\mathbf{R})$. Each loop in G_o is contractible in $Sp(n,\mathbf{R})$, hence $\rho^{-1}(G_0) \subseteq Mp(n,\mathbf{R})$ has two connected components. We denote by $\tilde{G}_o$ the connected component of $\rho^{-1}(G_o)$ containing the identity element. It is a normal subgroup of $\tilde{G}$ and the quotient group $\tilde{G}/\tilde{G}_o$ is isomorphic to $Mp(n-m,\mathbf{R})$, for $n > m$.

Let ξ be a Hamiltonian vector field in D_{12} and ϕ^t the one-parameter group of canonical transformations generated by ξ. Each ϕ^t induces a mapping $\phi^{\#t}: \mathscr{B}_\omega\mathscr{T}X \to \mathscr{B}_\omega\mathscr{T}X$, defined by

$$\phi^{\#t}(\underline{u};\underline{v}) = (\mathscr{T}\phi^t(u^1),\ldots,\mathscr{T}\phi^t(u^n);\mathscr{T}\phi^t(v^1),\ldots,\mathscr{T}\phi^t(v^n)), \quad (5.88)$$

which commutes with the right action of $Sp(n,\mathbb{R})$. Since ϕ^t preserves F_1 and F_2 it follows that $\phi^{\#t}$ maps $\mathscr{B}_\omega{}^{12}\mathscr{T}X$ onto itself. The one-parameter group $\phi^{\#t}$ lifts to a one-parameter group $\tilde{\phi}^{\#t}$ of automorphisms of the bundle $\tilde{\mathscr{B}}_\omega\mathscr{T}X$ of metaplectic frames. Since each $\phi^{\#t}$ maps $\mathscr{B}_\omega{}^{12}\mathscr{T}X$ onto itself, each $\tilde{\phi}^{\#t}$ maps $\tilde{\mathscr{B}}_\omega{}^{12}\mathscr{T}X$ onto itself.

The actions on $\mathscr{B}_\omega{}^{12}\mathscr{T}X$ of the one-parameter groups of canonical transformations generated by the Hamiltonian vector fields in D_{12} and of the group G_o lead to an equivalence relation $\sim$ in $\mathscr{B}_\omega{}^{12}\mathscr{T}X$. Given $\underline{b},\underline{b}' \in \mathscr{B}_\omega{}^{12}\mathscr{T}X$,

$$\underline{b} \sim \underline{b}' \Leftrightarrow \underline{b} = \phi^{\#t}(\underline{b}')g \quad (5.89)$$

for some one-parameter group ϕ^t of canonical transformations generated by a Hamiltonian vector field in D_{12} and some $g \in G_o$. The quotient space $\mathscr{B}_\omega{}^{12}\mathscr{T}X/\sim$ is a right principal $Sp(n-m,\mathbb{R})$ fibre bundle over X/D_{12}. The relation (5.89) can be lifted to $\tilde{\mathscr{B}}_\omega{}^{12}\mathscr{T}X$ as follows:

$$\tilde{\underline{b}} \sim \tilde{\underline{b}}' \Leftrightarrow \tilde{\underline{b}} = \tilde{\phi}^{\#t}(\tilde{\underline{b}}')\tilde{g} \quad (5.90)$$

for some one-parameter group ϕ^t of canonical transformations generated by a Hamiltonian vector field in D_{12} and some $\tilde{g} \in \tilde{G}_o$. The quotient space $\tilde{\mathscr{B}}_\omega{}^{12}\mathscr{T}X/\sim$ is a right principal $Mp(n-m,\mathbb{R})$ fibre bundle over X/D_{12} covering $\mathscr{B}_\omega{}^{12}\mathscr{T}X/\sim$.

For each integral manifold M of E_{12}, the restriction

of $\mathscr{B}_\omega{}^{12}\mathscr{T}X/\sim$ to $N = \pi_{D_{12}}(M)$ is isomorphic to the bundle $\mathscr{B}_{\omega_N}\mathscr{T}N$ of symplectic frames of (N,ω_N), and we shall identify these bundles. The restriction of $\tilde{\mathscr{B}}_\omega{}^{12}\mathscr{T}X/\sim$ to N defines a bundle $\tilde{\mathscr{B}}_{\omega_N}\mathscr{T}N$ of metaplectic frames for (N,ω_N). Let $\mathscr{P}_{\omega_N}\mathscr{T}N$ be the bundle of positive Lagrangian frames of (N,ω_N) and $\tilde{\mathscr{P}}_{\omega_N}\mathscr{T}N$ the bundle of metalinear positive Lagrangian frames of (N,ω_N) defined in terms of $\tilde{\mathscr{B}}_{\omega_N}\mathscr{T}N$. The bundles $\mathscr{P}_{\omega_N}\mathscr{T}N$ and $\tilde{\mathscr{P}}_{\omega_N}\mathscr{T}N$ can be represented as quotient spaces of subbundles of $\mathscr{P}_\omega\mathscr{T}X$ and $\tilde{\mathscr{P}}_\omega\mathscr{T}X$ respectively. Let $\mathscr{P}_\omega{}^{12}\mathscr{T}X$ be defined as follows:

$$\mathscr{P}_\omega{}^{12}\mathscr{T}X = \{(\underline{v},\underline{u}) \in \mathscr{P}_\omega\mathscr{T}X \mid \underline{v} \in \mathscr{B}D_{12}\} \tag{5.91}$$

and

$$\tilde{\mathscr{P}}_\omega{}^{12}\mathscr{T}X = \tau^{-1}(\mathscr{P}_\omega{}^{12}\mathscr{T}X). \tag{5.92}$$

We have a mapping $\upsilon\colon \mathscr{P}_\omega{}^{12}\mathscr{T}X|M \to \mathscr{P}_{\omega_N}\mathscr{T}N$ defined by

$$\upsilon(\underline{v},\underline{u}) = \mathscr{T}\pi_{12}(\underline{u}) \tag{5.93}$$

which lifts to $\tilde{\upsilon}\colon \tilde{\mathscr{P}}_\omega{}^{12}\mathscr{T}X|M \to \tilde{\mathscr{P}}_{\omega_N}\mathscr{T}N$.

We are now in a position to lift the conditions (5.23), (5.24) and (5.25) to the bundle of metalinear frames. First, we notice that the conditions (5.23) and (5.24) imply that $\underline{w}_1 = (\underline{v}_1,\underline{u}_1)$ and $\underline{w}_2 = (\underline{v}_1,\underline{u}_2)$ belong to $\mathscr{P}_\omega{}^{12}\mathscr{T}X$. The condition (5.25) can be extended to a relation in $\mathscr{P}_\omega{}^{12}\mathscr{T}X$ given by

$$i\omega(u_1{}^j,\bar{u}_2{}^k) = \hbar\,\delta^{jk}. \tag{5.94}$$

If M denotes the integral manifold of E_{12} passing through the point x at which the frames $\underline{w}_1$ and $\underline{w}_2$ are attached, we can write Eq. (5.94) in the form

$$i\omega_N(\mathscr{T}\pi_{12}(u_1{}^j),\ \mathscr{T}\pi_{12}(\bar{u}_2{}^k)) = \hbar\delta^{ik}. \tag{5.95}$$

Eq. (5.95) means that the projections $\upsilon(\underline{w}_1)$ and $\upsilon(\underline{w}_2)$ satisfy the relation (5.74) in $\mathscr{P}_{\omega_N}\mathscr{T}N$. We can restrict the choice of the metalinear frames $\underline{\tilde{w}}_1$ and $\underline{\tilde{w}}_2$ to be used in Eq. (5.29) by requiring that $\tilde{\upsilon}(\underline{\tilde{w}}_1)$ and $\tilde{\upsilon}(\underline{\tilde{w}}_2)$ be related in $\tilde{\mathscr{P}}_{\omega_N}\mathscr{T}N$ by the condition (5.80).

6. QUANTIZATION

We describe here the process of quantizing functions f on (X,ω) which generate one-parameter groups ϕ_f^t of canonical transformations such that the pair $(\mathscr{T}\phi_f^t(F),F)$ of polarizations is strongly admissible.

6.1. Lifting the action of ϕ_f^t

We have gathered together all the ingredients of the geometric quantization scheme. These are: a metaplectic frame bundle $\tilde{\mathscr{B}}_\omega\mathscr{T}X$, an associated bundle $\tilde{\mathscr{P}}_\omega\mathscr{T}X$ of positive metalinear Lagrangian frames, a complete strongly admissible positive polarization F, and a prequantization line bundle L with a connection satisfying Eq. (3.20) and an invariant Hermitian form. The metalinear frame bundle $\tilde{\mathscr{B}}F$ of F is defined by $\tilde{\mathscr{P}}_\omega\mathscr{T}X$ as follows:

$$\tilde{\mathscr{B}}F = \tau^{-1}(\mathscr{B}F), \tag{6.1}$$

where $\tau: \tilde{\mathscr{P}}_\omega\mathscr{T}X \to \mathscr{P}_\omega\mathscr{T}X$ is the double covering map. The representation space $\mathscr{U}$ consists of sections of $L \otimes \sqrt{\wedge^n F}$ covariantly constant along F, where $\sqrt{\wedge^n F}$ is associated to the bundle $\tilde{\mathscr{B}}F$ given by Eq. (6.1).

Consider a function f on X such that the Hamiltonian vector field ξ_f of f is complete, and let $\phi_f{}^t$ denote the one-parameter group of canonical transformations generated by ξ_f. For each $t \in \mathbf{R}$, the image of F under the derived mapping $\mathscr{T}\phi_f{}^t$ is a complete strongly admissible positive polarization $\mathscr{T}\phi_f{}^t(F)$ of (X,ω). The one-parameter group $\phi_f{}^t$ of transformations of (X,ω) lifts to a one-parameter group $\phi_f{}^{\#t}$ of diffeomorphisms of $\mathscr{P}_\omega\mathscr{T}X$ which preserves the structure of the bundle of positive Lagrangian frames: for each $\underline{w} \in (w_1,\ldots,w_n) \in \mathscr{P}_\omega\mathscr{T}X$, $\phi_f{}^{\#t}(\underline{w}) = (\mathscr{T}\phi_f{}^t(w_1),\ldots,\mathscr{T}\phi_f{}^t(w_n))$. There exists a unique lift of $\phi_f{}^{\#t}$ to a one-parameter group $\tilde{\phi}_f{}^{\#t}$ of structure preserving diffeomorphisms of $\tilde{\mathscr{P}}_\omega\mathscr{T}X$. For each $t \in \mathbf{R}$, the bundle $\tilde{\mathscr{P}}_\omega\mathscr{T}X$ induces a metalinear frame bundle of $\mathscr{T}\phi_f{}^t(F)$ given by $\tilde{\mathscr{B}}\mathscr{T}\phi_f{}^t(F) = \tau^{-1}(\mathscr{B}\mathscr{T}\phi_f{}^t(F))$ and, if $\underline{\tilde{w}} \in \tilde{\mathscr{B}}F$, then $\tilde{\phi}_f{}^{\#t}(\underline{\tilde{w}}) \in \tilde{\mathscr{B}}\mathscr{T}\phi_f{}^t(F)$. Thus, $\tilde{\phi}_f{}^{\#t}$ restricted to $\tilde{\mathscr{B}}F$ yields an isomorphism of the $ML(n,\mathbf{C})$-principal fibre bundles $\tilde{\mathscr{B}}F$ and $\tilde{\mathscr{B}}\mathscr{T}\phi_f{}^t(F)$. Let $\sqrt{\wedge^n}\mathscr{T}\phi_f{}^t(F)$ be the fibre bundle associated to $\tilde{\mathscr{B}}\mathscr{T}\phi_f{}^t(F)$ with typical fibre $\mathbf{C}$ on which $ML(n,\mathbf{C})$ acts by multiplication by $\chi(\underline{C})$, $\underline{C} \in ML(n,\mathbf{C})$. We denote by $\mathscr{H}_t$ the $\mathscr{T}\phi_f{}^t(F)$-representation space consisting of those sections of $L \otimes \sqrt{\wedge^n}\mathscr{T}\phi_f{}^t(F)$ which are covariantly constant along $\mathscr{T}\phi_f{}^t(F)$. If ν is a local section of $\sqrt{\wedge^n}F$, then for each $t \in \mathbf{R}$, we have a section $\phi_f{}^t\nu$ of $\sqrt{\wedge^n}\mathscr{T}\phi_f{}^t(F)$ defined by

$$(\phi_f{}^t\nu)^\#(\underline{\tilde{w}}) = \nu^\#(\tilde{\phi}_f{}^{\#-t}(\underline{\tilde{w}})) \tag{6.2}$$

for each $\underline{\tilde{w}} \in \tilde{\mathscr{B}}\mathscr{T}\phi_f{}^t(F)$. If ν is covariantly constant along F, then $\phi_f{}^t\nu$ is covariantly constant along $\mathscr{T}\phi_f{}^t(F)$.

For each element $\sigma \in \mathscr{H}$ which can be factorized,

$$\sigma = \lambda \otimes \nu, \tag{6.3}$$

we define $\phi_f{}^t\sigma$ by

$$\phi_f{}^t\sigma = \phi_f{}^t\lambda \otimes \phi_f{}^t\nu, \tag{6.4}$$

where $\phi_f{}^t\lambda$ is defined by Eq. (3.32). The definition (6.4) extends by linearity to all sections in $\mathscr{A}$. The map $\phi_f{}^t : \mathscr{A} \to \mathscr{A}_t$ is a vector space isomorphism with inverse defined in terms of $\phi_f{}^{-t}$. Moreover, for each $t \in \mathbf{R}$, the scalar product on $\mathscr{A}_t$ is defined intrinsically in terms of ω, L, $\tilde{\mathscr{P}}_\omega\mathscr{T}X$ and $\mathscr{T}\phi_f{}^t(F)$, cf. Chapter 4. Hence, $\phi_f{}^t : \mathscr{A} \to \mathscr{A}_t$ is a unitary map.

6.2. Polarization preserving functions

In this section we assume that

$$\mathscr{T}\phi_f{}^t(F) = F \tag{6.5}$$

for all $t \in \mathbf{R}$. This is equivalent to the condition that, for each complex vector field ξ in F, the Lie bracket $[\xi_f,\xi]$ be in F. In particular, if

$$\underline{\xi} = (\xi^1,\dots,\xi^n) \tag{6.6}$$

is a local frame field for F consisting of complex Hamiltonian vector fields defined in an open set $U \subseteq X$, there exists a matrix valued function $x \mapsto (a^i{}_j(x))$ on U such that

$$[\xi_f,\xi^i](x) = \sum_{j=1}^{n} a^i{}_j(x)\xi^j(x). \tag{6.7}$$

Eq. (6.5) implies that $\mathscr{A}_t = \mathscr{A}$ for all $t \in \mathbf{R}$. Hence, the function f generates a one-parameter group $\phi_f{}^t : \mathscr{A} \to \mathscr{A}$ of unitary transformations of $\mathscr{A}$. The *quantum operator* $\mathscr{Q}f$ on $\mathscr{A}$ associated to f is defined by

$$\mathscr{Q}f[\sigma] := i\hbar \frac{d}{dt}(\phi_f{}^t\sigma)\Big|_{t=0} \tag{6.8}$$

for each $\sigma \in \mathscr{A}$; it is a self-adjoint first-order differential operator.

We can give a local description of $\mathscr{Q}f$ as follows. Let $\underline{\tilde{\xi}}$ be a metalinear frame field on $U \subseteq X$ projecting onto $\underline{\xi}$ given by Eq. (6.6), and $\nu_{\underline{\tilde{\xi}}}$ the local section of $\sqrt{\wedge^n F}$ defined by

$$\nu_{\underline{\tilde{\xi}}} \circ \underline{\tilde{\xi}} = 1. \tag{6.9}$$

The restriction of σ to U can be factorized by $\nu_{\underline{\tilde{\xi}}}$,

$$\sigma|U = \lambda \otimes \nu_{\underline{\tilde{\xi}}} \tag{6.10}$$

for some covariantly constant section λ of $L|U$. We have

$$\begin{aligned}\mathscr{Q}f[\lambda \otimes \nu_{\underline{\tilde{\xi}}}] &= (i\hbar \frac{d}{dt}\, \phi_f{}^t\lambda)\Big|_{t=0} \otimes \nu_{\underline{\tilde{\xi}}} \\ &\quad + \lambda \otimes (i\hbar \frac{d}{dt}\, \phi_f{}^t\nu_{\underline{\tilde{\xi}}})\Big|_{t=0}.\end{aligned} \tag{6.11}$$

The first term on the right hand side is $\mathscr{P}f[\lambda] \otimes \nu_{\underline{\tilde{\xi}}}$, cf. Eq. (3.35). Thus it remains to evaluate the second term. Let $\phi_f{}^t\underline{\tilde{\xi}}$ denote the local metalinear frame field for F defined by

$$\phi_f{}^t\underline{\tilde{\xi}}(x) = \tilde{\phi}_f{}^{\#t}(\underline{\tilde{\xi}}(\phi_f{}^{-t}(x))). \tag{6.12}$$

The frames $\phi_f{}^t\underline{\tilde{\xi}}(x)$ and $\underline{\tilde{\xi}}(x)$ are related by an element $\tilde{C}_t(x) \in ML(n,\mathbf{C})$,

$$\phi_f{}^t\underline{\tilde{\xi}}(x) = \underline{\tilde{\xi}}(x)\tilde{C}_t(x). \tag{6.13}$$

We have

$$\begin{aligned}(\phi_f{}^t\nu_{\underline{\tilde{\xi}}})^{\#}(\underline{\tilde{\xi}}(x)) &= \nu_{\underline{\tilde{\xi}}}{}^{\#}(\tilde{\phi}_f{}^{\#-t}(\underline{\tilde{\xi}}(x))) \\ &= \nu_{\underline{\tilde{\xi}}}{}^{\#}(\phi_f{}^{-t}\underline{\tilde{\xi}}(\phi_f{}^{-t}(x))) \\ &= \nu_{\underline{\tilde{\xi}}}{}^{\#}(\underline{\tilde{\xi}}(\phi_f{}^{-t}(x))\tilde{C}_{-t}(\phi_f{}^{-t}(x)))\end{aligned}$$

$$= \chi\{[\tilde{C}_{-t}(\phi_f^{-t}(x))]^{-1}\}\nu_{\tilde{\underline{\xi}}}^{\#}(\tilde{\underline{\xi}}(\phi_f^{-t}(x)))$$

$$= \chi\{[\tilde{C}_{-t}(\phi_f^{-t}(x))]^{-1}\}\nu_{\tilde{\underline{\xi}}}^{\#}(\tilde{\underline{\xi}}(x)),$$

where the last equality follows from Eq. (6.9). Hence

$$\phi_f^{\,t}\nu_{\tilde{\underline{\xi}}}(x) = \chi\{[\tilde{C}_{-t}(\phi_f^{-t}(x))]^{-1}\}\nu_{\tilde{\underline{\xi}}}(x) \tag{6.14}$$

and

$$\frac{d}{dt}(\phi_f^{\,t}\nu_{\tilde{\underline{\xi}}}(x))\Big|_{t=0} = \frac{d}{dt}\,\chi\{[\tilde{C}_{-t}(\phi_f^{-t}(x))]^{-1}\}\Big|_{t=0}\nu_{\tilde{\underline{\xi}}}(x). \tag{6.15}$$

But $\tilde{C}_o(\phi_f^{\,t}(x)) = \underline{\tilde{I}}$, so that

$$\frac{d}{dt}\,\chi\{[\tilde{C}_{-t}(\phi_f^{-t}(x))]^{-1}\}\Big|_{t=0} = \frac{d}{dt}\,\chi\{[\tilde{C}_{-t}(x)]^{-1}\}\Big|_{t=0}. \tag{6.16}$$

Let $C_t(x)$ be the projection of $\tilde{C}_t(x)$ to $GL(n,R)$,

$$C_t(x) = \rho(\tilde{C}_t(x)). \tag{6.17}$$

Then

$$\phi_f^{\,t}\underline{\xi}(x) = \underline{\xi}(x)C_t(x) \tag{6.18}$$

and

$$\chi\{[\tilde{C}_{-t}(x)]^{-1}\} = \{\det[C_{\,t}(x)]^{-1}\}^{\frac{1}{2}}. \tag{6.19}$$

This implies that

$$\begin{aligned}\frac{d}{dt}\,\chi\{[\tilde{C}_{-t}(x)]^{-1}\}\Big|_{t=0} &= \frac{1}{2}\,\frac{d}{dt}\,\{\det[C_{-t}(x)]^{-1}\}\Big|_{t=0}\\ &= \frac{1}{2}\,\frac{d}{dt}\,[\det\,C_t(x)]\Big|_{t=0}\\ &= \frac{1}{2}\,\mathrm{tr}\,[\frac{d}{dt}\,C_t(x)]\Big|_{t=0},\end{aligned}$$

where tr denotes the trace of a matrix. Substituting this result into Eq. (6.15) we get

$$\frac{d}{dt}(\phi_f^{\,t}\nu_{\tilde{\underline{\xi}}}(x))\Big|_{t=0} = \frac{1}{2}\,\mathrm{tr}[\frac{d}{dt}\,C_t(x)]\Big|_{t=0}\nu_{\tilde{\underline{\xi}}}(x). \tag{6.20}$$

To evaluate the trace of the derivative of $C_t(x)$,

differentiate Eq. (6.18) with respect to t and set $t = 0$:

$$\underline{\xi}(x) \frac{d}{dt} C_t(x)\Big|_{t=0} = \frac{d}{dt}(\phi_f{}^t \underline{\xi}(x))\Big|_{t=0}$$

$$= \frac{d}{dt} (\phi_f{}^t \xi^1(x), \ldots, \phi_f{}^t \xi^n(x))\Big|_{t=0}$$

$$= (-[\xi_f, \xi^1](x), \ldots, -[\xi_f, \xi^n(x)])$$

$$= \Big(-\sum_{j=1}^{n} a^1{}_j(x)\xi^j(x), \ldots, -\sum_{j=1}^{n} a^n{}_j(x)\xi^j(x)\Big),$$

where the last equality follows from Eq. (6.7). Hence

$$\mathrm{tr}[\frac{d}{dt} C_t(x)]\Big|_{t=0} = -\sum_{j=1}^{n} a^j{}_j(x). \tag{6.21}$$

Substituting this result into Eq. (6.20) we get

$$\frac{d}{dt} \phi_f{}^t \nu_{\tilde{\underline{\xi}}}(x)\Big|_{t=0} = -\tfrac{1}{2}\sum_{j=1}^{n} a^j{}_j(x)\nu_{\tilde{\underline{\xi}}}(x). \tag{6.22}$$

This, together with Eqs. (6.11) and (3.35), yields

$$\mathcal{Q}f[\lambda \otimes \nu_{\tilde{\underline{\xi}}}] = \Big[\Big(\mathcal{P}f - \tfrac{1}{2}i\hbar \sum_{j=1}^{n} a^j{}_j\Big)\lambda\Big] \otimes \nu_{\tilde{\underline{\xi}}}. \tag{6.23}$$

Taking into account Eq. (3.38), we can rewrite Eq. (6.23) in the form

$$\mathcal{Q}f[\lambda \otimes \nu_{\tilde{\underline{\xi}}}] = \Big[\Big(-i\hbar\nabla_{\xi_f} + f - \tfrac{1}{2}i\hbar \sum_{j=1}^{n} a^j{}_j\Big)\lambda\Big] \otimes \nu_{\tilde{\underline{\xi}}}. \tag{6.24}$$

Eq. (6.24) can also be used to define $\mathcal{Q}f$ for a function f with incomplete Hamiltonian vector field ξ_f, in which case $\phi_f{}^t$ is only a local one-parameter group of local canonical transformations. However, in this circumstance the problem of the self-adjointness of $\mathcal{Q}f$ has to be studied separately.

Suppose that f is a function constant along F so that the Hamiltonian vector field ξ_f of f is contained in

F. Since the Hamiltonian vector fields contained in a polarization commute, the matrix $(a^i{}_j)$ defined by Eq. (6.7) vanishes identically. Moreover, for each local section λ of L covariantly constant along F, $\nabla_{\xi_f}\lambda = 0$. Thus, we can rewrite Eq. (6.24) in the form

$$\mathscr{Q}f[\lambda \otimes \nu_{\underline{\tilde{\xi}}}] = f\lambda \otimes \nu_{\underline{\tilde{\xi}}}. \tag{6.25}$$

Since the form of the right hand side of Eq. (6.25) is independent of the factorization (6.10), we have

$$\mathscr{Q}f[\sigma] = f\sigma \tag{6.26}$$

for each $\sigma \in \mathscr{A}$ and each function f constant along F. Hence, the operator $\mathscr{Q}f$ corresponding to a function f constant along F acts on quantum states σ by multiplication by the function f, and the spectrum of $\mathscr{Q}f$ is contained in the image $f(S)$ of the Bohr-Sommerfeld variety S under the function $f: X \to \mathbf{R}$.

Clearly, the mapping $f \mapsto \mathscr{Q}f$ given by Eq. (6.24) is a linear monomorphism from the space of polarization preserving functions to the space of symmetric linear operators on $\mathscr{A}$. Moreover, if f and g preserve F, then so does their Poisson bracket $[f,g]$. Using Eqs. (6.23), (6.7), (3.38), (3.1) and (2.10), we obtain by direct computation the commutation relations

$$[\mathscr{Q}f, \mathscr{Q}g] = i\hbar\mathscr{Q}[f,g]. \tag{6.27}$$

6.3. Quantization via Blattner-Kostant Sternberg kernels

We assume here that there exists an $\varepsilon > 0$ such that the pair $(F, \mathscr{T}\phi_f{}^t(F))$ of polarizations is strongly admissible for $0 < t < \varepsilon$, and that the integrals in Eqs. (5.12) and (5.22) defining the Blattner-Kostant-Sternberg kernel $\mathscr{K}_t: \mathscr{A}_t \times \mathscr{A} \to \mathbf{C}$

converge, so that there exists a linear map $\mathscr{U}_t : \mathscr{A}_t \to \mathscr{A}$ satisfying Eq. (5.1). For each $t \in (0,\varepsilon)$, we define $\Phi_t : \mathscr{A} \to \mathscr{A}$ by

$$\Phi_t = \mathscr{U}_t \circ \phi_f{}^t. \tag{6.28}$$

If the curve $t \mapsto \Phi_t$ is differentiable, we can define the quantum operator $\mathscr{Q}f$ corresponding to f by

$$\mathscr{Q}f := i\hbar \frac{d}{dt} \Phi_t \Big|_{t=0}. \tag{6.29}$$

If, for each $t \in (0,\varepsilon)$, the representation spaces $\mathscr{A}_t$ and $\mathscr{A}$ are *unitarily related* (that is, $\mathscr{U}_t$ is unitary), then ϕ_t is unitary, which implies that $\mathscr{Q}f$ is self-adjoint. In practice it may be difficult to verify the unitarity of $\mathscr{U}_t$, but one still may use Eq. (6.29) to compute $\mathscr{Q}f$ on a dense domain and investigate the existence of a self-adjoint extension afterwards.

In many applications to quantum mechanics, the polarization F is real, the leaves of D are simply connected, and the polarizations F and $\mathscr{T}\phi_f{}^t(F)$ are transverse, cf. Eq. (5.2). Under these conditions, one can give a useful expression for $\mathscr{Q}f[\sigma]$, where σ is a smooth section with sufficiently small support. Let V be a contractible open set in X/D such that on $\pi_D{}^{-1}(V)$ there exist n real-valued functions $q^1,\ldots,q^n$ such that their Hamiltonian vector fields $\xi_{q^1},\ldots,\xi_{q^n}$ span $F|\pi_D{}^{-1}(V)$. Let $p_1,\ldots,p_n$ be functions on $\pi_D{}^{-1}(V)$ such that

$$\theta := \sum_j p_j dq^j \tag{6.30}$$

satisfies

$$\omega|\pi_D{}^{-1}(V) = d\theta. \tag{6.31}$$

Then

$$\xi_{q^i} p_j = -\delta_{ij}, \quad i,j \in \{1,\ldots,n\}. \tag{6.32}$$

Let $\lambda_o : \pi_D^{-1}(V) \to L$ be a section such that

$$\nabla\lambda_o = -i\hbar^{-1}\theta \otimes \lambda_o \tag{6.33}$$

and

$$\langle\lambda_o, \lambda_o\rangle = 1. \tag{6.34}$$

We denote by $\underline{\xi}$ the linear frame field for $F|\pi_D^{-1}(V)$ consisting of the Hamiltonian vector fields of $q^1,\ldots,q^n$:

$$\underline{\xi} = (\xi_{q^1},\ldots,\xi_{q^n}), \tag{6.35}$$

and by $\underline{\tilde{\xi}}$ a metalinear frame field projecting onto $\underline{\xi}$. Let $\nu_{\underline{\tilde{\xi}}}$ be the section of $\sqrt{\wedge^n F}$ over $\pi_D^{-1}(V)$ defined by the condition

$$\nu_{\underline{\tilde{\xi}}}^{\#} \circ \underline{\tilde{\xi}} = 1. \tag{6.36}$$

Then, $\lambda_o \otimes \nu_{\underline{\tilde{\xi}}}$ is covariantly constant along F and each smooth section $\lambda \otimes \nu$ in $\mathscr{A}$, the support of which is contained in $\pi_D^{-1}(V)$, can be represented as

$$\lambda \otimes \nu = \psi(q^1,\ldots,q^n)\lambda_o \otimes \nu_{\underline{\tilde{\xi}}}, \tag{6.37}$$

where ψ is a smooth complex-valued function on R^n with support contained in the range of the mapping $(q^1,\ldots,q^n): \pi_D^{-1}(V) \to R^n$.

The functions $q^1,\ldots,q^n$ are constant along the fibres of π_D and they define coordinate functions $\check{q}^1,\ldots,\check{q}^n$ on V such that

$$q^i = \check{q}^i \circ \pi_D, \quad i = 1,\ldots,n. \tag{6.38}$$

The coordinate functions $\check{q}^i$, $i = 1,\ldots,n$, define a density

$$d^n\check{q} = |d\check{q}^1\wedge\ldots\wedge d\check{q}^n| \tag{6.39}$$

on V. The sections λ_o and $\nu_{\tilde{\underline{\xi}}}$ are chosen so that

$$\langle \lambda_o \otimes \nu_{\tilde{\underline{\xi}}}, \lambda_o \otimes \nu_{\tilde{\underline{\xi}}} \rangle = d^n \check{q}, \tag{6.40}$$

where the left hand side is defined by Eq. (4.26). The scalar product on $\mathscr{H}$ restricted to sections of the form (6.37) can be expressed as

$$\begin{aligned}(\psi'(q^1,\ldots,q^n)\lambda_o \otimes \nu_{\tilde{\underline{\xi}}} | \psi(q^1,\ldots,q^n)\lambda_o \otimes \nu_{\tilde{\underline{\xi}}}) \\ = \int_V \psi'(\check{q}^1,\ldots,\check{q}^n)\overline{\psi}(\check{q}^1,\ldots,\check{q}^n) d^n q.\end{aligned} \tag{6.41}$$

Taking into account Eqs. (5.1) and (6.28), we can write

$$\begin{aligned}&(\psi'(q^1,\ldots,q^n)\lambda_o \otimes \nu_{\tilde{\underline{\xi}}} | \Phi_t(\psi(q^1,\ldots,q^n)\lambda_o \otimes \nu_{\tilde{\underline{\xi}}})) \\ &= \mathscr{K}_t(\psi'(q^1,\ldots,q^n)\lambda_o \otimes \nu_{\tilde{\underline{\xi}}}, \psi(q^1\circ\phi_f^{-t},\ldots,q^n\circ\phi_f^{-t})\phi_f^{\,t}\lambda_o \otimes \phi_f^{\,t}\nu_{\tilde{\underline{\xi}}}),\end{aligned} \tag{6.42}$$

where $\phi_f^{\,t}\lambda_o$ is defined by Eq. (3.32) and $\phi_f^{\,t}\nu_{\tilde{\underline{\xi}}}$ is given by Eq. (6.2). Let ψ_t be defined by

$$\Phi_t(\psi(q^1,\ldots,q^n)\lambda_o \otimes \nu_{\tilde{\underline{\xi}}}) = \psi_t(q^1,\ldots,q^n)\lambda_o \otimes \nu_{\tilde{\underline{\xi}}}. \tag{6.43}$$

Substituting this into Eq. (6.42) we obtain, with the help of Eqs. (5.11), (5.12) and (6.41),

$$\begin{aligned}&\int_V \psi'(\check{q}^1,\ldots,\check{q}^n)\overline{\psi}_t(\check{q}^1,\ldots,\check{q}^n) d^n \check{q} \\ &= (i/\hbar)^{n/2} \int_{\pi_D^{-1}(V)} \Big\{ \Big[\det \omega(\xi_{q^j}, \phi_f^{\,t}\xi_{q^k})\Big]^{\frac{1}{2}} \\ &\quad \times \langle \lambda_o, \phi_f^{\,t}\lambda_o \rangle \psi'(q^1,\ldots,q^n)\overline{\psi}(q^1\circ\phi_f^{-t},\ldots,q^n\circ\phi_f^{-t})\Big\} d^n p\, d^n q,\end{aligned} \tag{6.44}$$

where

$$d^n p\, d^n q = |(\omega|\pi_D^{-1}(V))^n|. \tag{6.45}$$

Hence, for each point $y \in V$, we have

$$\psi_t(\check{q}^1(y),\ldots,\check{q}^n(y)) = (i\hbar)^{-n/2}\int_{\pi_D^{-1}(y)} \left\{\left[\det \omega(\xi_{q^j},\phi_f{}^t\xi_{q^k})\right]^{\frac{1}{2}} \times \langle\phi_f{}^t\lambda_o,\lambda_o\rangle\psi(q^1\circ\phi_f{}^{-t},\ldots,q^n\circ\phi_f{}^{-t})\right\}d^np_y, \tag{6.46}$$

where

$$d^np_y := |(dp_1\wedge\ldots\wedge dp_n)|\pi_D{}^{-1}(y)| \tag{6.47}$$

is a density on $\pi_D{}^{-1}(y)$.

Using the results of Sec. 3.3 we have

$$\begin{aligned}\frac{d}{dt}\langle\phi_f{}^t\lambda_o,\lambda_o\rangle &= (i\hbar)^{-1}\langle\phi_f{}^t(\mathscr{P}f[\lambda_o]),\lambda_o\rangle \\ &= (i\hbar)^{-1}\langle\phi_f{}^t((-i\hbar\nabla_{\xi_f}+f)\lambda_o),\lambda_o\rangle \\ &= (i\hbar)^{-1}\langle\phi_f{}^t((-\theta(\xi_f)+f)\lambda_o),\lambda_o\rangle \\ &= (i\hbar)^{-1}((-\theta(\xi_f)+f)\circ\phi_f{}^{-t})\langle\phi_f{}^t\lambda_o,\lambda_o\rangle.\end{aligned}$$

Therefore one has, by integration,

$$\langle\phi_f{}^t\lambda_o,\lambda_o\rangle = \exp\left\{(i/\hbar)\int_0^t [\theta(\xi_f)-f]\circ\phi_f{}^{-s}ds\right\}. \tag{6.48}$$

Substituting this result into Eq. (6.46) we find

$$\psi_t(\check{q}^1(y),\ldots,\check{q}^n(y)) = (i\hbar)^{-n/2}\int_{\pi_D^{-1}(y)} \left\{\left[\det \omega(\xi_{q^j},\phi_f{}^t\xi_{q^k})\right]^{\frac{1}{2}} \times \exp\left[(i/\hbar)\int_0^t(\theta(\xi_f)-f)\circ\phi_f{}^{-s}ds\right]\psi(q^1\circ\phi_f{}^{-t},\ldots,q^n\circ\phi_f{}^{-t})\right\}d^np_y. \tag{6.49}$$

According to Eqs. (6.29) and (6.43), we therefore have

$$\mathscr{Q}f[\psi\lambda_o\otimes\nu_{\underline{\tilde{\xi}}}] = i\hbar\left.\frac{d}{dt}\psi_t\right|_{t=0}\lambda_o\otimes\nu_{\underline{\tilde{\xi}}}, \tag{6.50}$$

where ψ_t is given by Eq. (6.49).

6.4. Superselection rules

Let F be a complete strongly admissible real polarization. In Sec. 4.5 we observed that to each connected component S_α of the Bohr-Sommerfeld variety S there corresponds a subspace $\mathscr{H}_\alpha$ of the representation space $\mathscr{H}$ consisting of sections with supports in S_α. The subspaces $\mathscr{H}_\alpha$ and $\mathscr{H}_\beta$ corresponding to different components S_α and S_β are orthogonal, and

$$\mathscr{H} = \oplus_\alpha \mathscr{H}_\alpha . \tag{6.51}$$

If f is a function on X preserving F then, for each component S_α of S, $\phi_f{}^t(S_\alpha) = S_\alpha$ and the operator $\mathscr{Q}f$ defined by Eq. (6.8) maps $\mathscr{H}_\alpha$ into itself. Suppose that f is a function such that $(F, \mathscr{T}\phi_f{}^t(F))$ is a strongly admissible pair of polarizations for $0 < t < \varepsilon$, and let σ be a section in $\mathscr{H}_\alpha$ such that $\pi_D(\text{support } \sigma)$ is compact. Then, for each component S_β of S different from S_α, there exists δ such that $0 < \delta < \varepsilon$ and

$$\phi_f{}^t(\text{support } \sigma) \cap S_\beta = \phi \tag{6.52}$$

for all $t \in (0,\delta)$. Hence,

$$(\Phi_t \sigma | \sigma') = 0 \tag{6.53}$$

for all $\sigma' \in \mathscr{H}_\beta$ and $t \in (0,\delta)$.

Differentiating Eq. (6.53) with respect to t, we obtain

$$(\mathscr{Q}f[\sigma] | \sigma') = 0 \tag{6.54}$$

for each $\sigma' \in \mathscr{H}_\beta$, where S_β is any component of S different from S_α. Since the space of sections σ in $\mathscr{H}_\alpha$, such that $\pi_D(\text{support } \sigma)$ is compact, is dense in $\mathscr{H}_\alpha$, it follows that $\mathscr{Q}f$ maps $\mathscr{H}_\alpha$ into itself. Hence, for each function f

which can be quantized according to the prescriptions given in this chapter, we have

$$\mathcal{Q}f[\mathcal{H}_\alpha] \subseteq \mathcal{H}_\alpha \tag{6.55}$$

for each connected component S_α of S.

Thus, if all the observables in the quantum theory described in terms of the representation space $\mathcal{H}$ are obtained by the quantization of functions on X in the manner described above, we obtain *superselection rules*. The representation space decomposes into a direct sum of orthogonal subspaces corresponding to different components of the Bohr-Sommerfeld variety, and all the observables commute with the operators of projection onto these subspaces.

7. SCHRÖDINGER REPRESENTATION

7.1. Single particle

The phase space X of a single particle is isomorphic to R^6. An isomorphism $X \to R^6$ is defined by the components q^1, q^2, q^3 of the position vector $\underline{q}$ and the components p_1, p_2, p_3 of the linear momentum $\underline{p}$ with respect to some inertial frame. The Lagrange bracket is given by

$$\omega = \sum_i dp_i \wedge dq^i. \tag{7.1}$$

Let

$$\theta = \sum_i p_i dq^i \tag{7.2}$$

so that $\omega = d\theta$. The *position representation* is given by the polarization F globally spanned by the linear frame field

$$\underline{\xi} = \left(\frac{\partial}{\partial p_1}, \frac{\partial}{\partial p_2}, \frac{\partial}{\partial p_3}\right). \tag{7.3}$$

Since X is contractible, all the additional structure needed for quantization exists and is unique up to isomorphism. Thus, the bundle $\tilde{\mathscr{B}}F$ of metalinear frames for F is trivial, and we denote by $\tilde{\underline{\xi}}$ a metalinear frame field for F projecting onto

$\underline{\xi}$. The section $\nu_{\underline{\tilde{\xi}}}$ of $\sqrt{\wedge^3 F}$ such that $\nu_{\underline{\tilde{\xi}}}^{\#} \circ \underline{\tilde{\xi}} = 1$ trivializes the bundle $\sqrt{\wedge^3 F}$. The bundle L is also trivial, $L = X \times \mathbf{C}$, with a trivializing section $\lambda_o: X \to L$ given by $\lambda_o(x) = (x,1)$ for each $x \in X$. The connection in L is defined by

$$\nabla\lambda_o = - i\hbar^{-1}\theta \otimes \lambda_o, \tag{7.4}$$

and the connection invariant Hermitian form is normalized by

$$\langle\lambda_o,\lambda_o\rangle = 1. \tag{7.5}$$

The representation space $\mathscr{H}$ consists of sections of $L \otimes \sqrt{\wedge^3 F}$ of the form

$$\sigma = \psi(\underline{q})\lambda_o \otimes \nu_{\underline{\tilde{\xi}}}, \tag{7.6}$$

where ψ is a square integrable complex function on $\mathbf{R}^3$.

The functions f on X of the form

$$f = \sum_i a^i(\underline{q})p_i + a(\underline{q}), \tag{7.7}$$

where a^1,a^2,a^3 and a are smooth real-valued functions on $\mathbf{R}^3$, generate canonical transformations $\phi_f^{\,t}$ which preserve F. We have

$$\xi_f = \sum_i\left(a^i\frac{\partial}{\partial q^i} - \frac{\partial a}{\partial q^i}\frac{\partial}{\partial p_i}\right) - \sum_{i,j}\frac{\partial a^i}{\partial q^j}p_i\frac{\partial}{\partial p_j}. \tag{7.8}$$

Hence,

$$\left[\xi_f, \frac{\partial}{\partial p_i}\right] = \sum_{i,j}\frac{\partial a^i}{\partial q^j}\frac{\partial}{\partial p_j} \tag{7.9}$$

and the matrix $(a^i{}_j)$ defined by Eq. (6.7) is given by

$$a^i{}_j = \frac{\partial a^i}{\partial q^j}. \tag{7.10}$$

Since $\nabla_{\xi_f}\lambda_o = (-i/\hbar)\theta(\xi_f)\lambda_o$, we obtain from Eq. (6.24) an expression for the action of $\mathscr{Q}f$ on $\lambda_o \otimes \nu_{\underline{\tilde{\xi}}}$:

$$\mathscr{Q}f\,[\lambda_o \otimes \nu_{\tilde{\underline{\xi}}}] = \Big(a - \tfrac{1}{2}i\hbar \sum_j \frac{\partial a^j}{\partial q^j}\Big)\lambda_o \otimes \nu_{\tilde{\underline{\xi}}}. \tag{7.11}$$

If ψ is a smooth function on R^3, the action of $\mathscr{Q}f$ on $\psi\lambda_o \otimes \nu_{\tilde{\underline{\xi}}}$ is

$$\mathscr{Q}f\,[\psi\lambda_o \otimes \nu_{\tilde{\underline{\xi}}}] = \Big[-i\hbar\xi_f\psi + \Big(a - \tfrac{1}{2}i\hbar \sum_j \frac{\partial a^j}{\partial q^j}\Big)\psi\Big]\lambda_o \otimes \nu_{\tilde{\underline{\xi}}}, \tag{7.12}$$

where ξ_f is given by Eq. (7.8). Of special importance in quantum theory are the components q^i of the position vector, the components p_i of the linear momentum vector and the components $J_i = \Sigma_{jk}\varepsilon_{ijk}q^j p_k$ of the angular momentum vector. Here, ε_{ijk} is the permutation symbol equaling 1 if $(1,2,3) \to (i,j,k)$ is an even permutation, -1 if it is an odd permutation and 0 if some of the indices i,j,k are equal.

In these cases, we obtain from Eq. (7.12) the standard results

$$\mathscr{Q}q^i\,[\psi\lambda_o \otimes \nu_{\tilde{\underline{\xi}}}] = (q^i\psi)\lambda_o \otimes \nu_{\tilde{\underline{\xi}}}, \tag{7.13}$$

$$\mathscr{Q}p^i\,[\psi\lambda_o \otimes \nu_{\tilde{\underline{\xi}}}] = \Big(-i\hbar\frac{\partial\psi}{\partial q^i}\Big)\lambda_o \otimes \nu_{\tilde{\underline{\xi}}} \tag{7.14}$$

and

$$\mathscr{Q}J_i\,[\psi\lambda_o \otimes \nu_{\tilde{\underline{\xi}}}] = \Big(-i\hbar\Sigma_{jk}\varepsilon_{ijk}q^j \frac{\partial\psi}{\partial q^k}\Big)\lambda_o \otimes \nu_{\tilde{\underline{\xi}}}. \tag{7.15}$$

Since the Hamiltonian vector fields of q^i, p_i and J_i are complete, the operators $\mathscr{Q}q^i$, $\mathscr{Q}p_i$ and $\mathscr{Q}J_i$ generate one-parameter groups of unitary transformations of $\mathscr{H}$.

The energy of a non-relativistic particle of mass m in an external potential $V(\underline{q})$ is given by

$$H = \underline{p}^2/2m + V(\underline{q}), \tag{7.16}$$

where

$$\underline{p}^2 = \sum_i p_i^{\,2}. \tag{7.17}$$

The Hamiltonian vector field of H is

$$\xi_H = \sum_i \left(m^{-1} p_i \frac{\partial}{\partial q^i} - \frac{\partial V}{\partial q^i} \frac{\partial}{\partial p_i} \right). \tag{7.18}$$

The local one-parameter group $\phi_H{}^t$ of canonical transformations generated by H does not preserve F. However, for a large class of potentials $\mathscr{T}\phi_H{}^t(F)$ is transverse to F if t is sufficiently small. We assume this transversality condition and proceed to quantize H in the manner of Sec. 6.3. From Eqs. (6.29) and (6.43) we have

$$\mathscr{Q}H\,[\psi\lambda_o \otimes \nu_{\tilde{\underline{\xi}}}] = i\hbar \frac{d}{dt}\psi_t\, \lambda_o \otimes \nu_{\tilde{\underline{\xi}}}\Big|_{t=0}, \tag{7.19}$$

where ψ_t is given by Eq. (6.49) which, in this case, takes the form

$$\psi_t(\underline{q}) = (i\hbar)^{-3/2} \int_{R^3} \left[\det \omega(\xi_{q^j}, \phi_H{}^t \xi_{q^k}) \right]^{\frac{1}{2}}$$
$$\times \exp\left\{ i\hbar^{-1} \int_0^t [\theta(\xi_H) - H] \circ \phi_H{}^{-s} ds \right\} \times \psi(\phi_H{}^t \underline{q})\, d^3 p. \tag{7.20}$$

Here $\phi_H{}^t \underline{q} = \underline{q} \circ \phi_H{}^{-t}$ is expressed as a function of $\underline{p}$ and q, and the integration is taken over the values of the components of the linear momentum $\underline{p}$ for a fixed position vector $\underline{q}$.

Since

$$\theta(\xi_H) - H = H - 2V(\underline{q}) = L \tag{7.21}$$

is the Lagrangian corresponding to the Hamiltonian H, and as H is invariant under $\phi_H{}^t$,

$$\begin{aligned} \int_0^t [\theta(\xi_H) - H] \circ \phi_H{}^{-s} ds &= tH - 2\int_0^t V(\phi_H{}^s \underline{q})\, ds \\ &= t\underline{p}^2/2m + tV(\underline{q}) - 2\int_0^t V(\phi_H{}^s \underline{q})\, ds. \end{aligned} \tag{7.22}$$

Substituting this expression back to Eq. (7.20), we obtain

$$\psi_t(\underset{\sim}{q}) = (ih)^{-3/2}\exp[it\hbar^{-1}V(\underset{\sim}{q})]\int_{R^3}\psi(\phi_H{}^t\underset{\sim}{q})\exp[-2i\hbar^{-1}\int_0^t V(\phi_H{}^s\underset{\sim}{q})ds]$$

$$\times \exp(it\underline{p}^2/2m\hbar)\{\det \omega(\xi^j,\phi_H{}^t\xi^k)\}^{\frac{1}{2}}d^3p, \tag{7.23}$$

where we have set $\xi^j = \frac{\partial}{\partial p_j}$. Changing the variables of integration from $\underline{p} = (p_1,p_2,p_3)$ to $\underline{u} = (u_1,u_2,u_3)$ given by

$$\underline{u} = t\underline{p}, \tag{7.24}$$

we get

$$\psi_t(\underset{\sim}{q}) = (ih)^{-3/2}\exp[it\hbar^{-1}V(\underset{\sim}{q})]\int_{R^3}\psi(\phi_H{}^t\underset{\sim}{q})\exp[-2i\hbar^{-1}\int_0^t V(\phi_H{}^s\underset{\sim}{q})ds]$$

$$\times t^{-3}\exp[i\underline{u}^2/2tm\hbar][\det \omega(\xi^j,\phi_H{}^t\xi^k)]^{\frac{1}{2}}d^3u. \tag{7.25}$$

The derivative of $\psi_t(\underset{\sim}{q})$ with respect to t can be evaluated at $t = 0$ by taking into account the facts that, for a real variable s and each $a > 0$,

$$\lim_{t\to 0+} t^{-\frac{1}{2}}\exp(ias^2/t) = (\pi/a)^{\frac{1}{2}}e^{-\pi/4}\delta(s) \tag{7.26}$$

where $\delta(s)$ is the Dirac distribution, and

$$\left(\frac{\partial^2}{\partial s^2} + 4ia\frac{\partial}{\partial t}\right)t^{-\frac{1}{2}}\exp(ias^2/t) = 0. \tag{7.27}$$

We shall see that the asymptotic behavior of $[\det \omega(\xi^j,\phi_H{}^t\xi^k)]^{\frac{1}{2}}$ as $t \to 0$ is given by $(t/m)^{3/2}$ plus higher-order terms. Hence, the second line in Eq. (7.25) leads to a distribution proportional to

$$\Delta\delta(\underline{u}) = \sum_i \frac{\partial^2}{\partial u_i{}^2}[\delta(u_1)\delta(u_2)\delta(u_3)]. \tag{7.28}$$

This implies that in order to evaluate the derivative of the right hand side of Eq. (7.25) with respect to t at $t = 0$ it suffices to approximate the integrand up to order 1 in t and order 2 in $\underline{u}$, and to neglect terms involving the product $t\underline{u}$ and higher-order terms. Since $\underline{u} = t\underline{p}$ it follows that it suffices to approximate the integrand in Eq. (7.23) up to order 1 in t and order 2 in $t\underline{p}$, and to omit terms involving $t^2\underline{p}$ and higher-order terms.

Eq. (7.18) yields the equations of motion

$$\frac{d}{dt}(q^i \circ \phi_H{}^t) = m^{-1} p_i \circ \phi_H{}^t \tag{7.29}$$

and

$$\frac{d}{dt}(p_i \circ \phi_H{}^t) = - \frac{\partial V}{\partial q^i}(q_i \circ \phi_H{}^t). \tag{7.30}$$

Hence,

$$\phi_H{}^t q^i = q^i \circ \phi_H{}^{-t} = q^i - m^{-1} t p_i + \text{higher-order terms.} \tag{7.31}$$

Moreover,

$$\phi_H{}^t \xi^i = \phi_H{}^t \xi_{q^i} = \xi_{\phi_H{}^t q^i} = \xi_{q^i \circ \phi_H{}^{-t}} \tag{7.32}$$

so that

$$\phi_H{}^t \xi^i = \xi^i - m^{-1} t \xi_{p_i} + \text{higher-order terms.} \tag{7.33}$$

Therefore,

$$\omega(\xi^j, \phi_H{}^t \xi^k) = m^{-1} t \delta^{jk} + \text{higher-order terms} \tag{7.34}$$

and

$$\{\det \omega(\xi^j, \phi_H{}^t \xi^k)\}^{\frac{1}{2}} = (t/m)^{3/2} + \text{higher-order terms.} \tag{7.35}$$

Also,

$$\int_0^t V(\phi_H{}^s \underline{q})\, ds = tV(\underline{q}) + \text{higher-order terms.} \tag{7.36}$$

Substituting the approximations given by Eqs. (7.31), (7.35) and (7.36) into Eq. (7.25) and taking into account Eq. (7.24), we find

$$\begin{aligned} \lim_{t\to 0+} \frac{d}{dt} \psi_t(\underline{q}) &= \lim_{t\to 0+} \frac{d}{dt}\{(-i/m\hbar)^{3/2} \exp[-it\hbar^{-1} V(\underline{q})] \\ &\quad \times \int_{R^3} \psi(\underline{q} - m^{-1}\underline{u}) t^{-3/2} \exp(i\underline{u}^2/2tm\hbar)\, d^3u\}, \end{aligned} \tag{7.37}$$

which leads with the help of Eqs. (7.26) and (7.27) to

$$\lim_{t \to 0+} \frac{d}{dt} \psi_t(\underset{\sim}{q}) = i\hbar^{-1} V(\underset{\sim}{q})\psi(\underset{\sim}{q}) + (i\hbar/2m)\Delta\psi(\underset{\sim}{q}). \tag{7.38}$$

Comparison of Eqs. (7.19) and (7.38) yields the standard expression for a quantized non-relativistic Hamiltonian

$$\mathscr{Q}H[\psi(\underset{\sim}{q})\lambda_o \otimes \nu_{\underset{\sim}{\xi}}] = \{[(-\hbar^2/2m)\Delta + V(\underset{\sim}{q})]\psi(\underset{\sim}{q})\}\lambda_o \otimes \nu_{\underset{\sim}{\xi}}. \tag{7.39}$$

Thus, we have recovered the correct operator expressions for the fundamental dynamical variables, cf. Eqs. (7.13), (7.14), (7.15) and (7.39). For each function on the phase space, which can be quantized via the Blattner-Kostant-Sternberg kernels, one obtains a well-defined operator and consequently there is no ambiguity in the ordering of operators.

7.2. System of particles

Let us consider a system of m interacting particles. The constraint part of the interaction defines an n-dimensional configuration manifold Y embedded in R^{3m}, and the kinetic energy of the system defines a positive definite metric g on Y. The phase space of the system is the cotangent bundle space $\mathscr{T}^*Y$ of Y and the Lagrange bracket is

$$\omega = d\theta_Y, \tag{7.40}$$

where θ_Y is the canonical 1-form on $\mathscr{T}^*Y$ defined by Eq. (2.12).

The dynamical variables of physical interest are the energy H, the canonical coordinates $q^1,\ldots,q^n$, and the corresponding canonical momenta $p_1,\ldots,p_n$. The canonical coordinates and momenta are defined on an open set $\pi^{-1}(U) \subseteq \mathscr{T}^*Y$, where U is the domain of a coordinate system on Y and

$\pi:\mathcal{T}^*Y \to Y$ is the cotangent bundle projection, cf. Sec. 2.2. Since the geometric quantization scheme applies to globally defined functions only, we have to redefine our notions of the canonical positions and momenta. The characteristic property of the canonical coordinates $q^1,\ldots,q^n$ is that they are constant along the fibres of the cotangent bundle projection. Indeed,

$$q^i = \check{q}^i \circ \pi, \quad i \in \{1,\ldots,n\} \tag{7.41}$$

where $(\check{q}^1,\ldots,\check{q}^n)$ are coordinate functions on Y. Thus, we say that a function f on $\mathcal{T}^*Y$ is of *position type* if it is constant along the fibres of $\pi:\mathcal{T}^*Y \to Y$. We shall denote the position type functions by q. For each smooth vector field ζ on Y we define a function p_ζ on $\mathcal{T}^*Y$ by

$$p_\zeta(x) = x(\zeta(\pi(x))) \quad \text{for each } x \in \mathcal{T}^*Y. \tag{7.42}$$

If $\zeta = \partial/\partial\check{q}^i$, then Eqs. (7.42) and (2.13) yield $p_\zeta(x) = p_i(x)$. For this reason we shall refer to p_ζ as the *momentum associated to a vector field* ζ. The Hamiltonian vector field of p_ζ projects onto ζ, i.e., for each $x \in \mathcal{T}^*Y$,

$$\mathcal{T}\pi(\xi_{p_\zeta}(x)) = \zeta(\pi(x)). \tag{7.43}$$

Using Eqs. (2.7), (2.8), (2.14) and (2.16) one can verify by direct computation that, for each pair (ζ,η) of smooth vector fields on Y and for each function $\check{q}$ on Y,

$$\theta_Y(\xi_{p_\zeta}) = p_\zeta \tag{7.44}$$

$$[\xi_{p_\zeta},\xi_{p_\eta}] = \xi_{p_{[\zeta,\eta]}} \tag{7.45}$$

$$[\xi_{p_\zeta},\xi_{\check{q}\circ\pi}] = \xi_{(\zeta\check{q})\circ\pi}. \tag{7.46}$$

Let D denote the *vertical distribution* on $\mathcal{T}^*Y$ defined by

$$D = \{v \in \mathcal{T}\mathcal{T}^*X \mid \mathcal{T}\pi(v) = 0\}. \tag{7.47}$$

The complexification F of D is a complete strongly admissible real polarization of $(\mathcal{T}^*Y, d\theta_Y)$. The space X/D of integral manifolds of D is diffeomorphic to Y, and we shall identify X/D with Y. The polarization F is locally spanned by the Hamiltonian vector fields of the canonical coordinate functions. The functions f of the form

$$f = \check{q}\circ\pi + p_\zeta, \tag{7.48}$$

where $\check{q}$ is a function on Y and ζ is a vector field on Y, generate the canonical transformations preserving F. Eqs. (7.45) and (7.46) imply that the functions f given by Eq. (7.48) form a subalgebra of the Poisson algebra of $(\mathcal{T}^*Y, d\theta_Y)$.

We assume that the configuration space Y is orientable. The choice of an orientation of Y induces a metaplectic structure on $(\mathcal{T}^*Y, d\theta_Y)$ which leads to a trivial bundle $\sqrt{\wedge^n F}$ as follows. Let $\{V_\alpha\}$ be a contractible covering of Y by domains of oriented charts $\check{q}_\alpha = (\check{q}_\alpha{}^1, \ldots, \check{q}_\alpha{}^n) : V_\alpha \to R^n$. We denote by $\underline{q}_\alpha = (q_\alpha{}^1, \ldots, q_\alpha{}^n)$ the pull-back of $\underline{\check{q}}^\alpha$ to $\pi^{-1}(V_\alpha)$, i.e., for each $i \in \{1, \ldots, n\}$

$$q_\alpha{}^i = \check{q}_\alpha{}^i\circ\pi, \tag{7.49}$$

and by $\underline{p}_\alpha = (p_{\alpha 1}, \ldots, p_{\alpha n})$ the map from $\pi^{-1}(V_\alpha)$ to R^n defined by

$$p_{\alpha i}(x) = x\left(\frac{\partial}{\partial \check{q}_\alpha{}^i}\bigg|_{\pi(x)}\right). \tag{7.50}$$

The ordered set of vector fields

$$(\xi_{\underline{q}_\alpha};\xi_{\underline{p}_\alpha}) = \left(\xi_{q_\alpha 1},\ldots,\xi_{q_\alpha n};\xi_{p_{\alpha 1}},\ldots,\xi_{p_{\alpha n}}\right) \tag{7.51}$$

on $\pi^{-1}(V_\alpha)$ forms a symplectic frame field such that

$$\xi_{\underline{q}_\alpha} = \left(\xi_{q_\alpha 1},\ldots,\xi_{q_\alpha n}\right) \tag{7.52}$$

is an oriented frame field for F. For each $x \in \pi^{-1}(V_\alpha \cap V_\beta)$ there exist $n \times n$ matrices $A_{\alpha\beta}(x)$ and $B_{\alpha\beta}(x)$ with

$$\det A_{\alpha\beta}(x) > 0 \tag{7.53}$$

such that

$$g_{\alpha\beta}(x) = \begin{pmatrix} A_{\alpha\beta}(x) & B_{\alpha\beta}(x) \\ 0 & [A_{\alpha\beta}{}^t(x)]^{-1} \end{pmatrix} \in Sp(n,\mathbf{R}) \tag{7.54}$$

and

$$(\xi_{\underline{q}_\beta}(x);\xi_{\underline{p}_\beta}(x)) = (\xi_{\underline{q}_\alpha}(x);\xi_{\underline{p}_\alpha}(x))g_{\alpha\beta}(x). \tag{7.55}$$

The collection G of matrices of the form (7.54) satisfying (7.53) constitutes a subgroup of $Sp(n,\mathbf{R})$ which can be continuously deformed in $Sp(n,\mathbf{R})$ to the identity element. Therefore, its preimage $\rho^{-1}(G)$ in $Mp(n,\mathbf{R})$ under the covering map $\rho\colon Mp(n,\mathbf{R}) \to Sp(n,\mathbf{R})$ has two connected components. Let $\tilde{G}$ denote the component of the identity element in $\rho^{-1}(G)$. The transition functions $g_{\alpha\beta}\colon \pi^{-1}(V_\alpha \cap V_\beta) \to G \subseteq Sp(n,\mathbf{R})$ have the unique lifts $\tilde{g}_{\alpha\beta}\colon \pi^{-1}(V_\alpha \cap V_\beta) \to \tilde{G} \subseteq Mp(n,\mathbf{R})$ defining a metaplectic frame bundle for $(\mathscr{T}^*Y, d\theta_Y)$ which induces a metalinear frame bundle $\tilde{\mathscr{B}}F$ for F. Let $\sqrt{\wedge^n F}$ denote the complex line bundle associated to $\tilde{\mathscr{B}}F$ corresponding to the character χ of $ML(n,\mathbf{C})$. Each $\pi^{-1}(V_\alpha)$ is contractible so that $\tilde{\mathscr{B}}F|\pi^{-1}(V_\alpha)$ is trivial and admits a section $\tilde{\xi}_{\underline{q}_\alpha}$ projecting onto $\xi_{\underline{q}_\alpha}$.

We denote by $\nu_\alpha : \pi^{-1}(V_\alpha) \to \sqrt{\wedge^n F}$ the local section such that

$$\nu_\alpha^{\#} \circ \tilde{\xi}_{\underline{q}_\alpha} = 1. \tag{7.56}$$

The metric g can be used to define a global nonvanishing section ν_g of $\sqrt{\wedge^n F}$. For each chart $(V_\alpha, \underline{q}_\alpha)$, g defines a matrix-valued function $g_\alpha = (g_{\alpha ij})$ on V_α such that

$$g | V_\alpha = \sum_{ij} g_{\alpha ij} d\check{q}_\alpha{}^i \otimes d\check{q}_\alpha{}^j. \tag{7.57}$$

Let $\underline{\zeta}_\alpha = (\zeta_{\alpha 1}, \ldots, \zeta_{\alpha n})$ be an oriented orthonormal frame field for $\mathscr{T}V_\alpha$. Expressing the vector fields $\zeta_{\alpha i}$ in terms of the frame field $(\partial/\partial\check{q}_\alpha{}^1, \ldots, \partial/\partial\check{q}_\alpha{}^n)$, we obtain

$$\zeta_{\alpha i} = \sum_j C_{\alpha i}{}^j \frac{\partial}{\partial \check{q}_\alpha{}^j}, \tag{7.58}$$

where $C_\alpha = (C_{\alpha i}{}^j)$ is a matrix-valued function on V_α such that

$$\det C_\alpha > 0 \tag{7.59}$$

and

$$(C_\alpha g_\alpha C_\alpha{}^t)_{ij} = \delta_{ij}. \tag{7.60}$$

It follows that

$$\det C_\alpha = |\det g_\alpha|^{-\frac{1}{2}}. \tag{7.61}$$

Let $\underline{\eta}_\alpha = (\eta_{\alpha 1}, \ldots, \eta_{\alpha n})$ be the frame field for $F | \pi^{-1}(V_\alpha)$ defined by

$$\eta_{\alpha i}\, p_{\zeta_{\alpha j}} = \delta_{ij}, \quad i,j \in \{1, \ldots, n\}. \tag{7.62}$$

Since $p_{\zeta_{\alpha j}} = \sum_k (C_{\alpha j}{}^k \circ \pi) p_{\alpha k}$, it follows that

$$\underline{\eta}_\alpha = \xi_{\underline{q}_\alpha} (C_\alpha{}^t \circ \pi)^{-1}. \tag{7.63}$$

Let $\tilde{\underline{\eta}}_\alpha$ be a metalinear frame field for F projecting onto $\underline{\eta}_\alpha$, and ν_{g_α} the local section of $\sqrt{\wedge^n F}$ defined by

$$\nu_{g_\alpha}^{\#} \circ \underline{\tilde{n}}_\alpha = 1. \tag{7.64}$$

Eqs. (4.12), (7.56), (7.63) and (7.64) yield

$$\nu_{g_\alpha} = \pm|(\det g_\alpha) \circ \pi|^{\frac{1}{4}}\nu_\alpha. \tag{7.65}$$

In $\pi^{-1}(V_\alpha \cap V_\beta)$ Eqs. (7.54) and (7.55) imply

$$dq^i_\beta = \sum_j dq_\alpha{}^j A_{\alpha\beta j}{}^i, \quad i \in \{1,\dots,n\}. \tag{7.66}$$

Hence, the matrix-valued function $A_{\alpha\beta} = (A_{\alpha\beta j}{}^i)$ is constant along the fibres of π, so that there exists a function $\check{A}_{\alpha\beta}$ on $V_\alpha \cap V_\beta$ such that

$$A_{\alpha\beta} = \check{A}_{\alpha\beta} \circ \pi. \tag{7.67}$$

Eqs. (7.57), (7.66) and (7.67) yield

$$g_\alpha = \check{A}_{\alpha\beta}{}^t g_\beta \check{A}_{\alpha\beta}. \tag{7.68}$$

Therefore, up to a factor ± 1, we have

$$\nu_{g_\alpha} = \pm|(\det g_\alpha) \circ \pi|^{\frac{1}{4}} = \pm|(\det g_\beta) \circ \pi|^{\frac{1}{4}}(\det A_{\alpha\beta})^{\frac{1}{2}}\nu_\alpha$$

$$= \pm|(\det g_\beta) \circ \pi|^{\frac{1}{4}}\nu_\beta = \pm\nu_{g_\beta}.$$

Since Y is orientable, there exists a global section ν_g of $\sqrt{\wedge^n F}$ such that, for each V_α,

$$\nu_g|\pi^{-1}(V_\alpha) = \pm\nu_{g_\alpha}. \tag{7.69}$$

Comparing with Eq. (7.65) we get

$$\nu_g|\pi^{-1}(V_\alpha) = \pm|(\det g_\alpha) \circ \pi|^{\frac{1}{4}}\nu_\alpha. \tag{7.70}$$

Since ν_α is covariantly constant along $F|\pi^{-1}(V_\alpha)$, it follows that ν_g is covariantly constant along F.

Since $\omega = d\theta_Y$ is exact, the prequantization line bundle L is trivial. We set $L = \mathscr{T}^*Y \times \mathbf{C}$, and define a trivializing section $\lambda_o: \mathscr{T}^*Y \to L$ by $\lambda_o(x) = (x,1)$. The connection ∇ in L satisfying the prequantization condition (3.20) is defined by

$$\nabla\lambda_o = -i\hbar^{-1}\theta_Y \otimes \lambda_o, \tag{7.71}$$

and the connection invariant Hermitian form is normalized by

$$\langle\lambda_o,\lambda_o\rangle = 1. \tag{7.72}$$

The section λ_o is covariantly constant along F, and each section σ of $L \otimes \sqrt{\wedge^n F}$ covariantly constant along F can be uniquely expressed in the form

$$\sigma = \Psi \otimes \lambda_o \otimes \nu_g, \tag{7.73}$$

where Ψ is a complex-valued function on Y. Since the integral manifolds of D are simply connected, the Bohr-Sommerfeld variety S coincides with the entire phase space $\mathscr{T}^*Y$, and the scalar product on the representation space $\mathscr{H}$, given by Eq. (4.28), reduces to

$$(\Psi_1 \otimes \lambda_o \otimes \nu_g|\Psi_2 \otimes \lambda_o \otimes \nu_g) = \int_Y \Psi_1\overline{\Psi}_2|\det g|^{\frac{1}{2}}. \tag{7.74}$$

Here, $|\det g|^{\frac{1}{2}}$ is the density on Y defined by

$$|\det g|^{\frac{1}{2}}\left(\frac{\partial}{\partial\check{q}_\alpha^1},\ldots,\frac{\partial}{\partial\check{q}_\alpha^n}\right) = |\det g_\alpha|^{\frac{1}{2}} \tag{7.75}$$

for each coordinate system $\underline{\check{q}}_\alpha: V_\alpha \to R^n$. In the derivation of Eq. (7.74), we have taken into account Eqs. (4.26), (7.56) and (7.65). Eq. (7.74) shows that the mapping $\mathscr{H} \to L^2(Y,|\det g|^{\frac{1}{2}}): \Psi \otimes \lambda_o \otimes \nu_g \mapsto \Psi$ is an isomorphism of the representation space $\mathscr{H}$ with the Hilbert space of complex-valued functions on Y square integrable with respect to the density $|\det g|^{\frac{1}{2}}$.

The quantization of functions of the form (7.48) proceeds via Eq. (6.24). For a function $q = \check{q}\circ\pi$, the corres- poinding operator is given by

$$\mathcal{Q}q[\Psi \otimes \lambda_o \otimes \nu_g] = (\check{q}\Psi) \otimes \lambda_o \otimes \nu_g. \tag{7.76}$$

If (support Ψ) $\subseteq V_\alpha$, the operator $\mathcal{Q}p_\zeta$ acting on $\Psi \otimes \lambda_o \otimes \nu_g$ can be expressed with the help of Eq. (7.65) as follows:

$$\begin{aligned}&\mathcal{Q}p_\zeta\Big[\Psi|\det g_\alpha|^{\frac{1}{4}} \otimes \lambda_o \otimes \nu_\alpha\Big] \\ &\quad = -i\hbar\Big\{[\zeta(\Psi|\det g_\alpha|^{\frac{1}{4}})]\circ\pi + \tfrac{1}{2}\sum_{j=1}^{n}(a^j{}_j)_\alpha[(\Psi|\det g_\alpha|^{\frac{1}{4}})\circ\pi]\Big\}\lambda_o \otimes \nu_\alpha,\end{aligned} \tag{7.77}$$

where the matrix $(a^i{}_j)_\alpha$ is defined in terms of the local frame field $\xi_{\underline{q}_\alpha}$ by Eq. (6.7). In order to compute $(a^i{}_j)_\alpha$ explicitly, let $(z_\alpha{}^1,\ldots,z_\alpha{}^n)$ be the components of ζ with respect to the frame field $\left(\frac{\partial}{\partial\check{q}_\alpha{}^1},\ldots,\frac{\partial}{\partial\check{q}_\alpha{}^n}\right)$,

$$\zeta|V_\alpha = \sum_j z_\alpha{}^j \frac{\partial}{\partial\check{q}_\alpha{}^j}. \tag{7.78}$$

Eqs. (7.46) and (7.49) yield

$$\Big[\xi_{p_\zeta},\xi_{q_\alpha{}^i}\Big] = \xi_{(\zeta\check{q}_\alpha{}^i)\circ\pi} = \xi_{z_\alpha{}^i\circ\pi} = \sum_j\left(\frac{\partial z_\alpha{}^i}{\partial\check{q}_\alpha{}^j}\circ\pi\right)\xi_{q_\alpha{}^j}.$$

Hence, $(a^i{}_j)_\alpha = \left(\frac{\partial z_\alpha{}^i}{\partial\check{q}_\alpha{}^j}\circ\pi\right)$ and

$$\begin{aligned}&\mathcal{Q}p_\zeta[\Psi \otimes \lambda_o \otimes \nu_g] \\ &\quad = -i\hbar|\det g_\alpha|^{-\frac{1}{4}}\Big[\sum_j\Big(z_\alpha{}^j\frac{\partial}{\partial\check{q}_\alpha{}^j} + \tfrac{1}{2}\frac{\partial z_\alpha{}^j}{\partial\check{q}_\alpha{}^j}\Big)(\Psi|\det g_\alpha|^{\frac{1}{4}})\Big] \otimes \lambda_o \otimes \nu_g.\end{aligned} \tag{7.79}$$

The term in the square bracket is a local expression for the Lie derivative of the $\frac{1}{2}$-density $\Psi|\det g|^{\frac{1}{4}}$ with respect to the vector field ζ,

$$[\mathscr{L}_\zeta(\Psi|\det g|^{\frac{1}{4}})]\left(\frac{\partial}{\partial \check{q}_\alpha^{\,1}},\ldots,\frac{\partial}{\partial \check{q}_\alpha^{\,n}}\right)$$
$$= \sum_j \left(z_\alpha^{\,j}\frac{\partial}{\partial \check{q}_\alpha^{\,j}} + \tfrac{1}{2}\frac{\partial z_\alpha^{\,j}}{\partial \check{q}_\alpha^{\,j}}\right)(\Psi|\det g_\alpha|^{\frac{1}{4}}). \tag{7.80}$$

This can be written as follows:

$$\mathscr{L}_\zeta(\Psi|\det g|^{\frac{1}{4}}) = |\det g|^{\frac{1}{4}}[\zeta\Psi + \tfrac{1}{2}(\mathrm{Div}\ \zeta)\Psi], \tag{7.81}$$

where $\mathrm{Div}\ \zeta$ is the covariant divergence of the vector field ζ. Thus

$$\mathscr{Q}p_\zeta[\Psi\otimes\lambda_0\otimes\nu_g] = -i\hbar[\zeta\Psi + \tfrac{1}{2}(\mathrm{Div}\ \zeta)\Psi]\otimes\lambda_0\otimes\nu_g. \tag{7.82}$$

We see that the operator of the momentum associated to a vector field ζ corresponds to $-i\hbar$ times the operator of differentiation in the direction ζ if and only if

$$\mathrm{Div}\ \zeta = 0. \tag{7.83}$$

Eq. (7.83) is equivalent to the condition that the local one-parameter group generated by ζ should preserve the metric density $|\det g|^{\frac{1}{2}}$. This holds in particular when ζ is a Killing vector field, that is is, $\mathscr{L}_\zeta g = 0$.

We denote by $K(x)$ the kinetic energy of a state $x \in \mathscr{T}^*Y$; it is determined by

$$K(x) = \tfrac{1}{2}\, g(x,x). \tag{7.84}$$

If the potential energy of the system is given by a function $V\colon Y \to \mathbf{R}$, the total energy is

$$H(x) = K(x) + V(\pi(x)). \tag{7.85}$$

The quantization of H proceeds in a manner analogous to that employed in the previous section. We have

$$\mathscr{Q}H[\Psi \otimes \lambda_o \otimes \nu_g] = i\hbar \frac{d}{dt} \Phi_t(\Psi \otimes \lambda_o \otimes \nu_g)\Big|_{t=0}. \tag{7.86}$$

If the support of Ψ is compact and contained in some coordinate neighborhood V_α, we can write

$$\Psi \otimes \lambda_o \otimes \nu_g = \psi(\underline{q}_\alpha)\lambda_o \otimes \nu_\alpha, \tag{7.87}$$

where ψ is a function on R^n with compact support contained in the image of the chart $\check{\underline{q}}_\alpha : V_\alpha \to R^n$. Comparing Eqs. (7.87) and (7.70) we find, for each $y \in V_\alpha$,

$$\psi(\check{\underline{q}}_\alpha(y)) = \pm\Psi(y)\,|\det g_\alpha(y)|^{\frac{1}{4}}. \tag{7.88}$$

Similarly, for sufficiently small $t \in (0,\varepsilon)$,

$$\Phi_t(\Psi \otimes \lambda_o \otimes \nu_g) = \psi_t(\underline{q}_\alpha)\lambda_o \otimes \nu_\alpha \tag{7.89}$$

where ψ_t is given by Eq. (6.49).

In the following we work with a fixed coordinate system $(\pi^{-1}(V_\alpha), \underline{q}_\alpha, \underline{p}_\alpha)$ on $\mathscr{T}^*Y$. In order to simplify the notation we drop the index α distinguishing the various coordinate systems. Thus we write (q,p) instead of (q_α, p_α), g_{ij} instead of $g_{\alpha ij}$, etc. With this notation we can write ψ_t in the form

$$\begin{aligned}\psi_t(\check{\underline{q}}(y)) = (ih)^{-n/2}\int_{\mathscr{T}_y^*Y} \psi(\phi_H{}^t \underline{q})\left[\det \omega(\xi_{q^j}, \phi_H{}^t \xi_{q^k})\right]^{\frac{1}{2}} \\ \times \exp\left\{i\hbar^{-1}\int_0^t [\theta_Y(\xi_H) - H] \circ \phi_H{}^{-s}\, ds\right\} d^n p_y.\end{aligned} \tag{7.90}$$

To evaluate the derivative of Φ_t at $t = 0$ we approximate the right hand side of Eq. (7.90) up to the first order in t. As in Sec. 7.1 we have, if $x \in \mathscr{T}_y^*Y$,

$$\int_0^t [\theta(\xi_H)-H] \circ \phi_H{}^{-s}(x)ds = \int_0^t (H-2V\circ\pi) \circ \phi_H{}^{-s}(x)ds$$

$$= tH(x) - 2\int_0^t V(\pi(\phi_H{}^{-s}(x))ds$$

$$= \frac{t}{2}\sum_{ij} g^{ij}(y)p_ip_j + tV(y) - 2\int_0^t V(\pi(\phi_H{}^{-s}(x))ds$$

$$= \frac{t}{2}\sum_{ij} g^{ij}(y)p_ip_j - tV(y) + \text{higher-order terms.}$$

Moreover, Eqs. (7.31) and (7.33) imply that the approximations needed in the first two factors under the integral sign in Eq. (7.90) are insensitive to the choice of the potential V. Consequently, in these terms we can replace $\phi_H{}^t$ by the one-parameter group $\phi_K{}^t$ generated by the kinetic energy K. Hence,

$$\psi_t(\underset{\sim}{\check{q}}(y)) = (i\hbar)^{-n/2} \exp\{-it\hbar^{-1}V(y)\}\int_{\mathcal{T}_y{}^*Y} \psi(\phi_K{}^t\underset{\sim}{q})$$

$$\times \left[\det \omega(\xi_{q^j}, \phi_K{}^t\xi_{q^k})\right]^{\frac{1}{2}}\{\exp(it/2\hbar)\sum_{jk} g^{jk}(y)p_jp_k\}d^np_y \quad (7.91)$$

$$+ \text{higher-order terms.}$$

The kinetic energy $K(x)$ is $\frac{1}{2}$ times the square of the covector x. Hence, the orbits of the one-parameter group $\phi_K{}^t$ project onto geodesics in Y. Computations analogous to those used in Sec. 2.3 with $N = 2K$ and $e = 0$ yield

$$\frac{d}{dt}(q^i\circ\phi_K{}^t) = \sum_j [(g^{ij}\circ\pi)p_j]\circ\phi_K{}^t \quad (7.92)$$

and

$$\frac{d}{dt}(p_i\circ\phi_K{}^t) = -\tfrac{1}{2}\sum_{jk} [(g^{jk}{}_{,i}\circ\pi)p_jp_k]\circ\phi_K{}^t. \quad (7.93)$$

Here, $g^{jk}{}_{,i}$ denotes $\partial g^{jk}/\partial\check{q}^i$. Differentiating Eq. (7.92) with respect to t and taking Eq. (7.93) into account we obtain

$$\frac{d^2}{dt^2}(q^i \circ \phi_K{}^t) = -\sum_{mn}(\Gamma_{mn}{}^i \circ \pi \circ \phi_K{}^t)\frac{d}{dt}(q^m \circ \phi_K{}^t)\frac{d}{dt}(q^n \circ \phi_K{}^t), \quad (7.94)$$

where the Christoffel symbols $\Gamma_{mn}{}^i$ are given by Eq. (2.30). The integral in Eq. (7.91) can be simplified if the coordinates $\check{q}^1,\ldots,\check{q}^n$ are normal at y. In this case the functions q^i depend linearly on the parameter t along the geodesics originating at y,

$$q^i \circ \phi_K{}^t(x) = tq^i \circ \phi_K{}^1(x) \quad (7.95)$$

for each $x \in \mathcal{T}_y{}^*Y$, and

$$g_{ij}(y) = \delta_{ij}. \quad (7.96)$$

Eqs. (7.94) and (7.95) imply that the Christoffel symbols of the metric g vanish at y,

$$\Gamma_{mn}{}^i(y) = 0, \quad (7.97)$$

which is equivalent to the vanishing at y of the first derivatives of the components g_{ij} of the metric tensor. Differentiating Eq. (7.94) we get, with the help of Eqs. (7.95) and (7.97),

$$\Gamma_{ij}{}^m{}_{,k}(y) + \Gamma_{ki}{}^m{}_{,j}(y) + \Gamma_{jk}{}^m{}_{,i}(y) = 0. \quad (7.98)$$

Here, $\Gamma_{ij}{}^m{}_{,k}$ is the derivative of $\Gamma_{ij}{}^m$ with respect to $\check{q}^k$.

Using Eqs. (7.92) and (7.93) we approximate the integrand in Eq. (7.91) so that the integration will give results accurate to first order in t. For this, it suffices to approximate the integrand up to order 1 in t and order 2 in tp_i and to omit terms of order t^2p_i and higher. Eq. (7.92) yields

$$q^i \circ \phi_K^{-t}(x) = q^i(x) - t\sum_j g^{ij}(\pi(x))p_j(x) \qquad (7.99)$$
$$+ \text{ higher-order terms.}$$

Since the $\check{q}^i$ are normal at y, we get, for each $x \in \mathscr{T}_y^*Y$,

$$q^i \circ \phi_K^{-t}(x) = -t \sum_j g^{ij}(y)p_j(x). \qquad (7.100)$$

Moreover,

$$\phi_K{}^t \xi_{q^j} = \xi_{q^j \circ \phi_K^{-t}} \qquad (7.101)$$

so that

$$\omega(\xi_{q^j}(x), \phi_K{}^t \xi_{q^k}(x)) = \{\xi_{q^j}[q^k - t \sum_i (g^{ki} \circ \pi)p_i]\}(x) \qquad (7.102)$$
$$+ \text{ higher-order terms} = t\, g^{jk}(y) + \text{ higher-order terms.}$$

Taking into account Eq. (7.96), we thus have

$$\left[\det \omega(\xi_{q^j}, \phi_K{}^t \xi_{q^k})\right]^{1/2} = t^{n/2} + \text{ higher-order terms.} \qquad (7.103)$$

Substituting Eqs. (7.96), (7.100), and (7.103) into Eq. (7.91), we obtain

$$\psi_t(\underline{0}) = (ih)^{-n/2} \exp[-it\hbar^{-1}V(y)]$$
$$\times\, t^{n/2} \int_{R^n} \psi(-t\underline{p}) \exp[(it/2\hbar)\sum_j p_j{}^2] d^n p \qquad (7.104)$$
$$+ \text{ higher-order terms.}$$

Changing the variables of integration from p_j to

$$u_j = tp_j, \qquad (7.105)$$

Eq. (7.104) becomes

$$\psi_t(\underline{0}) = (ih)^{-n/2} \exp[-it\hbar^{-1}V(g)]$$
$$\times\, t^{-n/2} \int_{R^n} \psi(-\underline{u}) \exp[(i/2t\hbar)\sum_j u_j{}^2] d^n u \qquad (7.106)$$
$$+ \text{ higher-order terms.}$$

Taking into account Eqs. (7.26) and (7.27) we obtain

$$\lim_{t\to 0+} \frac{d}{dt}\psi_t(\underline{0}) = -i\hbar^{-1}V(y)\psi(\underline{0}) + (i\hbar/2)\sum_m \frac{\partial^2\psi}{\partial u_m^2}(\underline{0}). \quad (7.107)$$

Eqs. (7.88), (7.95), (7.96) and (7.97) yield

$$\sum_m \frac{\partial^2\psi}{\partial u_m^2}(\underline{0}) = \pm\sum_{mn} g^{mn}(y)\frac{\partial}{\partial\check{q}^m}\frac{\partial}{\partial\check{q}^n}[\Psi|\det g|^{\frac{1}{4}}](y) \quad (7.108)$$

where, according to our convention, we have suppressed the subscript α designating the chosen coordinate system. Performing the differentiation of the product on the right hand side of Eq. (7.108) and using Eqs. (7.97) and (7.98), we can rewrite Eq. (7.108) in the explicitly covariant form

$$\sum_m \frac{\partial^2\psi}{\partial u_m^2}(\underline{0}) = \pm[\det g(y)]^{\frac{1}{4}}\{\Delta\Psi(y) - \tfrac{1}{6}R(y)\Psi(y)\}. \quad (7.109)$$

In Eq. (7.109) Δ denotes the Laplace-Beltrami operator given by

$$\Delta\Psi = \sum_{mn} g^{mn}\,\nabla_m\nabla_n\Psi, \quad (7.110)$$

where ∇_m denotes covariant differentiation in the direction $\partial/\partial\check{q}^m$, and R is the scalar curvature of the metric connection in Y defined by

$$R = \sum_{imn} g^{mn}R^i{}_{mni}, \quad (7.111)$$

where

$$R^k{}_{mni} = \Gamma_{mn}{}^k{}_{,i} - \Gamma_{mi}{}^k{}_{,n} + \sum_j(\Gamma_{mn}{}^j\Gamma_{ji}{}^k - \Gamma_{mi}{}^k\Gamma_{jn}{}^k). \quad (7.112)$$

Differentiating Eq. (7.89) with respect to t and setting $t = 0$ we obtain, with the help of Eqs. (7.87), (7.88), (7.107) and (7.109)

$$\frac{d}{dt}\Phi_t(\Psi\otimes\lambda_o\otimes\nu_g)\Big|_{t=0} = \{-i\hbar^{-1}V\Psi + \frac{i\hbar}{2}[\Delta\Psi - \frac{1}{6}R\Psi]\}\otimes\lambda_o\otimes\nu_g. \tag{7.113}$$

Since $\mathscr{Q}H = i\hbar\frac{d}{dt}\Phi_t\Big|_{t=0}$ we have finally

$$\mathscr{Q}H[\Psi\otimes\lambda_o\otimes\nu_g] = \{-\frac{\hbar^2}{2}[\Delta\Psi - \frac{1}{6}R\Psi] + V\Psi\}\otimes\lambda_o\otimes\nu_g. \tag{7.114}$$

As in the case of a single particle, we have obtained operator expressions for the basic dynamical variables of the theory, cf. Eqs. (7.76), (7.82) and (7.114). The expression (7.114) for the energy contains an additional term proportional to the scalar curvature of the configuration space. It should be noted that the validity of the result for $\mathscr{Q}H$ obtained here depends on the convergence of the integrals defining the Blattner-Kostant-Sternberg kernels, and this poses certain restrictions on the geometry of the configuration space.

7.3. Blattner-Kostant-Sternberg kernels, quasiclassical approximation, and Feynman path integrals

In the previous sections of this chapter we obtained the quantum operators corresponding to all the fundamental dynamical variables of the theory with the exception of the square of the angular momentum vector. Among these dynamical variables only the energy required the Blattner-Kostant-Sternberg kernels for its quantization. In this section we want to examine the relation of the Blattner-Kostant-Sternberg kernels to both the quasiclassical approximation and quantization by means of the Feynman path integrals.

Let us look back at the technique employed in the

quantization of the energy. Eqs. (7.86) and (7.89) yield

$$\mathcal{Q}H[\psi\,\lambda_0 \otimes \nu_\alpha] = i\hbar \frac{d}{dt}\psi_t\,\lambda_0 \otimes \nu_\alpha\Big|_{t=0}, \tag{7.115}$$

where ψ_t is given by Eq. (7.90). Changing the integration variables from $\underline{p}|\mathcal{T}_y^*Y$ to

$$\underline{q}_0 = \phi_H{}^t\underline{q}|\mathcal{T}_y^*Y, \tag{7.116}$$

we can rewrite Eq. (7.90) in the form

$$\psi_t(\underline{q}) = \int_{R^n} d^nq_0\psi(\underline{q}_0)K(\underline{q}_0;t,\underline{q}), \tag{7.117}$$

where

$$K(\underline{q}_0;t,\underline{q}) := (i\hbar)^{-n/2}\Big[\det\omega(\xi_{q^j},\phi_H{}^t\xi_{q^k})\Big]^{\frac{1}{2}}$$

$$\times\exp\Big\{i\hbar^{-1}\int_0^t[\theta_Y(\xi_H)-H]\circ\phi_H{}^{-s}ds\Big\}\Big[\det\Big(\frac{\partial p_j}{\partial q_0{}^k}\Big)\Big]. \tag{7.118}$$

The integral in the exponent in Eq. (7.118) is the action $S(0,\underline{q}_0;t,\underline{q})$ along the trajectory of the system starting at $\underline{q}_0 = q\circ\phi_H{}^{-t}$ at time 0 and ending at $\underline{q}$ at time t:

$$\int_0^t[\theta_Y(\xi_H)-H]\circ\phi_H{}^{-s}ds = S(0,\underline{q}_0;t,\underline{q}). \tag{7.119}$$

This result becomes obvious when one notices that

$$L = \theta_Y(\xi_H) - H \tag{7.120}$$

is the Lagrangian corresponding to the Hamiltonian H. Eq. (7.119) can also be obtained by direct computation from Eq. (2.53) if one takes into account Eqs. (2.45), (2.50) and (2.51) with $\tilde{H}$ replaced by the time-independent Hamiltonian H discussed here.

Since $\xi_{q^i} = -\frac{\partial}{\partial p_i}$ and $\phi_H{}^t\xi_{q^k} = \xi_{\phi_H{}^t q^k}$, we have on the fibre $\mathcal{T}_y^*Y$

$$\det \omega(\xi_{q^j}, \phi_H{}^t \xi_{q^k}) = \det\left[\left(\xi_{\phi_H{}^t q^k} \rfloor \omega\right)\left(\frac{\partial}{\partial p_j}\right)\right]$$

$$= \det\left(-\frac{\partial \phi_H{}^t q^k}{\partial p_j}\right)$$

$$= \det\left(-\frac{\partial q_o{}^k}{\partial p_j}\right)$$

$$= (-1)^n \left[\det\left(\frac{\partial p_j}{\partial q_o{}^k}\right)\right]^{-1}.$$

But Eq. (2.54) yields

$$p_j = \frac{\partial}{\partial q_j} S(0,\underline{q}_o;t,\underline{q}); \tag{7.121}$$

hence

$$\det \omega(\xi_{q^j}, \phi_H{}^t \xi_{q^k}) = (-1)^n \left[\det\left(\frac{\partial^2 S}{\partial q^j \partial q_o{}^k}\right)\right]^{-1}. \tag{7.122}$$

Similarly, the Jacobian of the change of integration variables is given by

$$\det\left(\frac{\partial p_j}{\partial q_o{}^k}\right) = \det\left(\frac{\partial^2 S}{\partial q^j \partial q_o{}^k}\right). \tag{7.123}$$

Substituting Eqs. (7.119), (7.122) and (7.123) into Eq. (7.118) we obtain

$$K(\underline{q}_o;t,\underline{q}) = (-i\hbar)^{-n/2} \left\{\det\left[\frac{\partial^2 S(0,\underline{q}_o;t,\underline{q})}{\partial q^j \partial q_o{}^k}\right]\right\}^{\frac{1}{2}} \times \exp[i\hbar^{-1} S(0,\underline{q}_0;t,\underline{q})]. \tag{7.124}$$

We see that, apart from the factor $(-i\hbar)^{-n/2}$, the right hand side of Eq. (7.124) corresponds to the *quasiclassical approximation* to the time-dependent Schrödinger equation: the exponential factor gives the W.K.B. approximation, and the Van Vleck determinant is a solution of the corresponding transport equation. Thus, the one-parameter family $\Phi_t : \mathscr{A} \to \mathscr{A}$

defined by the Blattner-Kostant-Sternberg kernels corresponds to the quasiclassical evolution of states. Each Φ_t is an integral operator with kernel $K(\underline{q}_o;t,\underline{q})$ given by Eq. (7.124),

$$\Phi_t(\psi(\underline{q})\lambda_o \otimes \nu_\alpha) = \left\{\int_{R^n} d^n q_o \ \psi(\underline{q}_o)K(\underline{q}_o;t,\underline{q})\right\}\lambda_o \otimes \nu_\alpha. \quad (7.125)$$

The quantum Hamiltonian $\mathscr{Q}H$ is given by Eq. (7.86), that is,

$$\mathscr{Q}H[\psi(\underline{q})\lambda_o \otimes \nu_\alpha] = i\hbar \frac{d}{dt}\Phi_t(\psi(\underline{q})\lambda_o \otimes \nu_\alpha)\Big|_{t=0}. \quad (7.126)$$

Let $\exp(-it\hbar^{-1}\mathscr{Q}H)$ denote the one-parameter group of unitary transformations of $\mathscr{H}$ generated by $\mathscr{Q}H$. For each smooth function ψ with compact support in R^n, the curves in $\mathscr{H}$ given by $t \mapsto \exp(-it\hbar^{-1}\mathscr{Q}H)[\psi(\underline{q})\lambda_o \otimes \nu_\alpha]$ and $t \mapsto \Phi_t[\psi(\underline{q})\lambda_o \otimes \nu_\alpha]$ are tangent at $t = 0$. Therefore, by interating Φ_t, one can express $\exp(-it\hbar^{-1}\mathscr{Q}H)$ as follows. Divide the interval $[0,t]$ into N equal subintervals. For large N, $\exp(-i(t/N)\hbar^{-1}\mathscr{Q}H)[\psi(\underline{q})\lambda_o \otimes \nu_\alpha]$ can be approximated by $\Phi_{t/N}[\psi(\underline{q})\lambda_o \otimes \nu_\alpha]$ and $\exp(-it\hbar^{-1}\mathscr{Q}H)[\psi(\underline{q})\lambda_o \otimes \nu_\alpha]$ can be approximated by

$$\begin{aligned}[\Phi_{t/N}]^N[\psi(\underline{q})\lambda_o \otimes \nu_\alpha] &= \left\{\int_{R^{nN}} d^n q_o \ \ldots \ d^n q_{N-1}\psi(\underline{q}_o)\right. \\ &\quad \left. \times K(\underline{q}_o;t/N,\underline{q}_1) \times \ldots \times K(\underline{q}_{N-1};t/N,\underline{q})\right\}\lambda_o \otimes \nu_\alpha.\end{aligned} \quad (7.127)$$

If the left hand side of Eq. (7.127) converges to $\exp(-it\hbar^{-1}\mathscr{Q}H)\psi(\underline{q})\lambda_o \otimes \nu_\alpha$ as $N \to \infty$, then Eq. (7.124) yields , for each $x \in \mathscr{T}^*Y$,

$$[\exp(-i\hbar^{-1}t\mathscr{Q}H)\psi(\underline{q})\lambda_o\otimes\nu_\alpha](x) = \lim_{N\to\infty}\Big\{(-i\hbar)^{-nN/2}\int_{R^{nN}} d^nq_o\ldots d^nq_{N-1}\psi(\underline{q}_o)$$

$$\times\left[\prod_{r=0}^{N-1}\det\left(\frac{\partial^2 S(0,\underline{q}_r;t/N,\underline{q}_{r+1})}{\partial q_r{}^j\partial q^k{}_{r+1}}\right)\right]^{\frac{1}{2}} \tag{7.128}$$

$$\times\exp\left[i\hbar^{-1}\sum_{r=0}^{N-1}S(0,\underline{q}_r;t/n,\underline{q}_{r+1})\right]\Big\}\lambda_o(x)\otimes\nu_\alpha(x),$$

where $\underline{q}_N = \underline{q}(x)$.

For a fixed point $x \in X$ and fixed values of the integration variables $\underline{q}_1,\underline{q}_2,\ldots,\underline{q}_{N-1}$, let $\gamma\colon [0,t] \to X$ be the piece-wise continuous curve defined by the conditions

$$\underline{q}(\gamma(rt/n)) = \underline{q}_r \tag{7.129}$$

and

$$\gamma(-s+rt/n) = \phi_H{}^{-s}(\gamma(rt/n)) \tag{7.130}$$

for each $r \in \{0,1,\ldots,N\}$ and each $s \in [0,t/N]$. Eqs. (7.119), (7.120), (7.129) and (7.130) yield

$$\begin{aligned}
&\sum_{r=0}^{N-1} S(0,\underline{q}_r;t/N,\underline{q}_{r+1})\\
&= \sum_{r=0}^{N-1} S(0,\phi_H{}^{t/N}\underline{q}(\gamma(r+1)t/N));t/N,\underline{q}(\gamma((r+1)t/N)))\\
&= \sum_{r=0}^{N-1}\int_0^{t/N} L\circ\phi_H{}^{-s}(\gamma((r+1)t/N))ds\\
&= \sum_{r=0}^{N-1}\int_0^{t/N} L(\gamma(-s+(r+1)t/N))ds\\
&= \sum_{r=0}^{N-1}\int_{rt/N}^{(r+1)t/N} L(\gamma(s))ds = \int_0^t L(\gamma(s))ds.
\end{aligned}$$

On the other hand, the Van Vleck determinant can be approximated for large N by $C_N t^{-nN/2}$, cf. Eqs. (7.103) and (7.122). Here, C_N is a normalization constant depending on the kinetic energy of the system, e.g., for the single particle discussed in Sec. 7.1 we would have $C_N = m^{3N/2}$. Thus, Eq. (7.128) can be rewritten, for each $x \in \mathcal{T}_y^*Y$, in the *Feynman path integral* form

$$[\exp(-i\hbar^{-1}t\mathcal{D}H)\psi(\underline{q})\lambda_o \otimes \nu_\alpha](x) = \left\{\int \mathcal{D}[\check{\gamma}]\psi(\underline{q}(\gamma(0)))\exp\left[i\hbar^{-1}\int_0^t L(\gamma(s))ds\right]\right\}\lambda_o(x) \otimes \nu_\alpha(x), \tag{7.131}$$

where γ is a path in X such that $\underline{q}(\gamma(t)) = \underline{q}(x)$, $\check{\gamma}$ is the projection of γ onto Y, $\check{\gamma} = \pi\circ\gamma$, and $\mathcal{D}[\check{\gamma}]$ is the Feynman pseudomeasure [cf. Eq. (7.140)] on the space of paths $\check{\gamma}$: $[0,t] \to Y$ satisfying

$$\check{\gamma}(t) = y. \tag{7.132}$$

The results of the preceding paragraph are valid only in the domain of the coordinate system $(\underline{p},\underline{q})$ on $\mathcal{T}^*Y$. In particular, they are applicable globally if and only if the coordinates $\check{\underline{q}}$ on Y define a diffeomorphism of Y with R^n. To globalize these results we must rewrite their derivation in terms of geometrically defined objects. Eqs. (7.87), (7.88), (7.89) and (7.117) yield

$$\Phi_t(\Psi\otimes\lambda_o\otimes\nu_g) = \left\{\int_{R^n} d^nq_o\Psi(\underline{q}_o)|\det g_\alpha(\underline{q}_o)|^{\frac{1}{4}}K(\underline{q}_o;t,\underline{q})\right\} \times |\det g_\alpha(\underline{q})|^{-\frac{1}{4}} \otimes \lambda_o \otimes \nu_g, \tag{7.133}$$

where we have put $\underline{q}_o$ as the argument of Ψ to represent the point in Y corresponding to the coordinate $\underline{q}_o$. Since $d^nq_o|\det g_\alpha(\underline{q}_o)|^{\frac{1}{2}}$ corresponds to the density $|\det g|^{\frac{1}{2}}$ on Y [cf. Eq. (7.75)], we have, for each $x \in \mathcal{T}_y^*Y$,

$$\Phi_t(\Psi \otimes \lambda_0 \otimes \nu_g)(x)$$
$$= \left\{\int_{y_0 \in Y} |\det g(y_0)|^{\frac{1}{2}}\Psi(y_0)K_t(y_0,y)\right\} \otimes \lambda_0(x) \otimes \nu_g(x) \tag{7.134}$$

where the integration is taken with respect to the argument y_0. $K_t(y_0,y)$ is a two-point function, that is, a function on a neighborhood of the diagonal in $Y \times Y$, and is defined by

$$K_t(y_0,y) = |\det g(y_0)|^{-\frac{1}{4}}K(\underline{q}(y_0);t,\underline{q}(y))|\det g(y)|^{-\frac{1}{4}}, \tag{7.135}$$

where the function K is given by Eq. (7.118). Taking into account Eq. (7.124) we can write

$$K_t(y_0,y)$$
$$= (-i\hbar)^{-n/2}|\det g(y_0)|^{-\frac{1}{4}}[\det D_yD_{y_0}S_0{}^t(y_0,y)]^{\frac{1}{2}}|\det g(y)|^{-\frac{1}{4}}$$
$$\times \exp\{(i\hbar^{-1}S_0{}^t(y_0,y)\}, \tag{7.136}$$

where $S_0{}^t(y_0,y)$ is the action integral defined by Eq. (2.45) and related to $S(0,\underline{q}_0;t,\underline{q})$ by Eq. (2.53), and

$$\det D_yD_{y_0}S_0{}^t(y_0,y)$$
$$= \det\left[\left.\frac{\partial^2 S(0,\underline{q}_0;t,\underline{q})}{\partial q^j \partial q_0{}^k}\right|_{\underline{q}_0=\underline{q}(y_0),\ \underline{q}=\underline{q}(y)}\right]. \tag{7.137}$$

The function $S_0{}^t(y_0,y)$ is a scalar two-point function, so that $D_yD_{y_0}S_0{}^t(y_0,y)$ is a two-point covector field. Therefore, under coordinate transformations $K_t(y_0,y)$ transforms as a scalar two-point function on Y. Hence, Eq. (7.134) expresses Φ_t as an integral operator with a scalar kernel $K_t(y_0,y)$ given by Eq. (7.136), where the integration is performed with respect to the metric density $|\det g|^{\frac{1}{2}}$.

To obtain an expression for the Feynman path integral we have to iterate Eq. (7.134). Instead of Eq. (7.127) we obtain, for $x \in \mathcal{T}_y^*Y$,

$$[\Phi_{t/N}]^N(\Psi \otimes \lambda_o \otimes \nu_g)(x)$$

$$= \left\{\int_{(y_o,\dots,y_{N-1})\in Y^N} |\det g(y_o)|^{\frac{1}{2}} \times \dots \times |\det g(y_{N-1})|^{\frac{1}{2}}\right.$$

$$\left. \times\ \Psi(y_o)K_{t/N}(y_o,y_1)\times\dots\times K_{t/N}(y_{N-1},y)\right\}\otimes\lambda_o(x)\otimes\nu_g(x). \quad (7.138)$$

Following the same steps which led to Eq. (7.131) we obtain

$$[\exp(-i\hbar^{-1}t\,\mathscr{D}H)\Psi \otimes \lambda_o \otimes \nu_g](x)$$

$$= \left\{\int \mathscr{D}_g[\check{\gamma}]\Psi(\gamma(0))\exp\left[i\hbar^{-1}\int_0^t L(\gamma(s))ds\right]\right\}\otimes\lambda_o(x)\otimes\nu_g(x), \quad (7.139)$$

where $\check{\gamma}$ is a path in Y satisfying Eq. (7.132). The curve $\gamma\colon [0,t] \to \mathcal{T}^*Y$ is obtained from $\check{\gamma}$ by the Legendre transformation, and $\mathscr{D}_g[\check{\gamma}]$ is a pseudomeasure on the space of paths satisfying Eq. (7.132) which can be formally expressed as

$$\mathscr{D}_g[\check{\gamma}] = \lim_{N\to\infty} (-i\hbar)^{-nN/2} \prod_{r=0}^{N-1} \left\{[\det g(\gamma(rt/N))]^{\frac{1}{4}}\right.$$

$$\left. \times\ [\det D_{\check{\gamma}((r+1)t/N)}D_{\check{\gamma}(rt/N)}S_o^{t/N}(\check{\gamma}(rt/N),\check{\gamma}((r+1)t/N))]^{\frac{1}{2}}\right\}$$

$$\times\ [\det g(y)]^{-\frac{1}{4}}. \quad (7.140)$$

8. OTHER REPRESENTATIONS

The choice of representation is determined by the choice of polarization. The polarization spanned by the Hamiltonian vector fields of the position functions gives rise to the Schrödinger representation. The momentum representation corresponds to the polarization spanned by the Hamiltonian vector fields of the momentum variables. The Blattner-Kostant-Sternberg kernel between these representations reduces to the Fourier transform. In this chapter, we describe the Bargmann-Fock representation defined by the polarization spanned by the Hamiltonian vector fields of complex coordinates on the phase space as well as the harmonic oscillator energy representation.

8.1. Bargmann-Fock representation

Let (X,ω) be the symplectic manifold introduced in Sec. 7.1 which represents the phase space of a single particle. The *Bargmann-Fock representation* is given by the polarization F_B spanned by the Hamiltonian vector fields of the complex coordinates z_k defined by

$$z_k = 2^{-\frac{1}{2}}(p_k + iq^k). \tag{8.1}$$

Denoting by $\bar{z}_k$ the complex conjugate of z_k,

$$\bar{z}_k = 2^{-\frac{1}{2}}(p_k - iq^k), \tag{8.2}$$

we can write

$$\omega = -i \sum_k d\bar{z}_k \wedge dz_k. \tag{8.3}$$

The Hamiltonian vector fields of the coordinates z_k can then be expressed as

$$\xi_{z_k} = -i \frac{\partial}{\partial \bar{z}_k} = -i2^{-\frac{1}{2}}\left(\frac{\partial}{\partial p_k} + i \frac{\partial}{\partial q^k}\right). \tag{8.4}$$

We have

$$i\omega(\xi_{z_k}, \bar{\xi}_{z_k}) = 1, \tag{8.5}$$

so that the polarization F_B spanned by the frame field

$$\xi_{\underline{z}} = (\xi_{z_1}, \xi_{z_2}, \xi_{z_3}) \tag{8.6}$$

is positive. Since

$$F_B \cap \bar{F}_B = 0 \quad \text{and} \quad F_B + \bar{F}_B = \mathscr{T}^{\mathbf{C}}X, \tag{8.7}$$

the polarization F_B is obviously strongly admissible and complete, cf. Sec. 4.1. The contractability of X implies that the bundle $\tilde{\mathscr{B}}F_B$ of metalinear frames for F_B is trivial. We denote by $\tilde{\xi}_{\underline{z}}$ a metalinear frame field for F_B covering $\xi_{\underline{z}}$ and by $\nu_{\tilde{\xi}_{\underline{z}}}$ the section of $\sqrt{\wedge^3 F_B}$ defined by

$$\nu_{\tilde{\xi}_{\underline{z}}}^{\#} \circ \tilde{\xi}_{\underline{z}} = 1. \tag{8.8}$$

Let $(L, \nabla, \langle , \rangle)$ be the prequantization structure introduced in Sec. 7.1. That is, L is the trivial line bundle, $L = X \times \mathbf{C}$, with a trivializing section $\lambda_o: X \to L: x \mapsto (x,1)$ such that

$$\nabla\lambda_o = -i\hbar^{-1}(\sum_k p_k dq^k) \otimes \lambda_o \tag{8.9}$$

and

$$<\lambda_o,\lambda_o> = 1. \tag{8.10}$$

We introduce a new trivializing section λ_1 of L defined by

$$\lambda_1 = \exp\{-(4\hbar)^{-1} \sum_k [(q^k)^2+(p_k)^2-2ip_kq^k]\}\lambda_o. \tag{8.11}$$

By direct computation we obtain

$$\nabla\lambda_1 = -i\hbar^{-1}\,\theta_1 \otimes \lambda_1, \tag{8.12}$$

where

$$\theta_1 = -i \sum_k \bar{z}_k dz_k. \tag{8.13}$$

Eqs. (8.12) and (8.13) imply that λ_1 is covariantly constant along F_B. Hence, every section σ of $L \otimes \sqrt{\wedge^3 F_B}$ which is covariantly constant along F_B can be expressed in the form

$$\sigma = \Psi(\underline{z})\lambda_1 \otimes \nu_{\tilde{\xi}_{\underline{z}}}, \tag{8.14}$$

where Ψ is a holomorphic function of the complex coordinates $\underline{z} = (z_1,z_2,z_3)$.

We denote by $\mathcal{H}_B$ the representation space defined by the polarization F_B. Since $D_B{}^{\mathbf{C}} = F_B \cap \bar{F}_B = 0$, the Bohr-Sommerfeld variety S is equal to the entire phase space X and $X/D_B = X$. Therefore the scalar product on $\mathcal{H}_B$ is given by Eqs. (4.28) and (4.26), and can be written in the form

$$(\Psi_1\lambda_1\otimes\nu_{\tilde{\xi}_{\underline{z}}}|\Psi_2\lambda_1\otimes\nu_{\tilde{\xi}_{\underline{z}}}) = \int_{\mathbf{R}^6} \Psi_1(\underline{z})\bar{\Psi}_2(\underline{z})<\lambda_1,\lambda_1>d^3p\,d^3q. \tag{8.15}$$

Taking into account Eqs. (8.11) and (8.10), we see that

$$<\lambda_1,\lambda_1> = \exp\{-(2\hbar)^{-1} \sum_k [(q^k)^2 + (p_k)^2]\} \tag{8.16}$$

which can be written as

$$\langle\lambda_1,\lambda_1\rangle = \exp(-|\underline{z}|^2/\hbar), \tag{8.17}$$

where

$$|\underline{z}|^2 = \sum_k \bar{z}_k z_k. \tag{8.18}$$

Hence, we have the following expression for the scalar product on $\mathscr{H}_B$:

$$(\Psi_1\lambda_1\otimes\nu_{\tilde{\xi}_{\underline{z}}}|\Psi_2\lambda_1\otimes\nu_{\tilde{\xi}_{\underline{z}}}) = \int_{R^6}\Psi_1(\underline{z})\bar{\Psi}_2(\underline{z})\exp(-|\underline{z}|^2/\hbar)d^3pd^3q. \tag{8.19}$$

Eqs. (8.14) and (8.19) establish an isomorphism of the representation space $\mathscr{H}_B$ with the space of holomorphic functions Ψ on C^3 with the scalar product given by the right hand side of Eq. (8.19).

As in the case of the Schrödinger representation, the components of the position, the momentum, and the angular momentum vectors preserve the polarization. Hence, they are quantized in the manner described in Sec. 6.2. In terms of the complex variables z_k and $\bar{z}_k$ defined by Eqs. (8.1) and (8.2), we have

$$p_k = 2^{-\frac{1}{2}}(z_k + \bar{z}_k) \tag{8.20}$$

$$q^k = -i2^{-\frac{1}{2}}(z_k - \bar{z}_k) \tag{8.21}$$

and

$$J_k = i\sum_{jm}\varepsilon_{kjm}z_j\bar{z}_m. \tag{8.22}$$

Using the expression (8.3) for ω we can write

$$\xi_{p_k} = i2^{-\frac{1}{2}}\left(\frac{\partial}{\partial z_k} - \frac{\partial}{\partial\bar{z}_k}\right) \tag{8.23}$$

$$\xi_{q^k} = -2^{-\frac{1}{2}}\left(\frac{\partial}{\partial z_k} + \frac{\partial}{\partial\bar{z}_k}\right) \tag{8.24}$$

and

$$\xi_{J_k} = -\sum_{jm} \varepsilon_{kjm}(z_j \frac{\partial}{\partial z_m} - \bar{z}_m \frac{\partial}{\partial \bar{z}_j}). \tag{8.25}$$

Substituting Eqs. (8.23), (8.24) and (8.25) into Eq. (6.24), we obtain

$$\mathcal{Q}_B p_k [\Psi\lambda_1 \otimes \nu_{\tilde{\xi}_{\underline{z}}}] = 2^{-\frac{1}{2}}[(z_k + \hbar \frac{\partial}{\partial z_k})\Psi]\lambda_1 \otimes \nu_{\tilde{\xi}_{\underline{z}}} \tag{8.26}$$

$$\mathcal{Q}_B q^k [\Psi\lambda_1 \otimes \nu_{\tilde{\xi}_{\underline{z}}}] = -i2^{-\frac{1}{2}}[(z_k - \hbar \frac{\partial}{\partial z_k})\Psi]\lambda_1 \otimes \nu_{\tilde{\xi}_{\underline{z}}} \tag{8.27}$$

$$\mathcal{Q}_B J_k [\Psi\lambda_1 \otimes \nu_{\tilde{\xi}_{\underline{z}}}] = i\hbar \sum_{jm} \varepsilon_{kjm} z_j \frac{\partial}{\partial z_m} \lambda_1 \otimes \nu_{\tilde{\xi}_{\underline{z}}}, \tag{8.28}$$

where we have used the symbol $\mathcal{Q}_B$ to denote quantization in the Bargmann-Fock representation.

The Bargmann-Fock representation is particularly convenient for quantizing the harmonic oscillator. Let H be the Hamiltonian of a harmonic oscillator with unit mass and spring constant,

$$H = \tfrac{1}{2} \sum_k [(q^k)^2 + (p_k)^2]. \tag{8.29}$$

In terms of the complex coordinates z_k and $\bar{z}_k$ we have

$$H = \sum_k \bar{z}_k z_k. \tag{8.30}$$

Therefore,

$$\xi_H = i\sum_k (z_k \frac{\partial}{\partial z_k} - \bar{z}_k \frac{\partial}{\partial \bar{z}_k}) \tag{8.31}$$

and

$$[\xi_H, \xi_{z_k}] = i\xi_{z_k}. \tag{8.32}$$

Thus, ξ_H preserves the polarization F_B and the matrix $a^j{}_k$ defined by Eq. (6.7) is

$$a^j{}_k = i\delta^j{}_k. \tag{8.33}$$

Substituting Eqs. (8.31) and (8.33) into Eq. (6.24) we obtain

$$\mathscr{Q}_B H[\Psi\lambda_1 \otimes \nu_{\tilde{\xi}_{\underline{z}}}] = \hbar(\sum_k z_k \frac{\partial\Psi}{\partial z_k} + \frac{3}{2}\Psi)\lambda_1 \otimes \nu_{\tilde{\xi}_{\underline{z}}}. \tag{8.34}$$

Let us consider an eigenstate of $\mathscr{Q}_B H$ with eigenvalue E. Eq. (8.34) yields

$$\hbar(\sum_k z_k \frac{\partial\Psi}{\partial z_k} + \frac{3}{2}\Psi) = E\Psi,$$

which is equivalent to

$$\sum_k z_k \frac{\partial\Psi}{\partial z_k} = (E/\hbar - \tfrac{3}{2})\Psi. \tag{8.35}$$

Eq. (8.35) implies that Ψ is a homogeneous function of z_k of degree $(E/\hbar - \frac{3}{2})$. Since Ψ is holomorphic it follows that $(E/\hbar - \frac{3}{2})$ is a non-negative integer n so that the spectrum of $\mathscr{Q}_B H$ is given by $E_n = (n + \frac{3}{2})\hbar$ for $n = 0,1,2,\ldots$. It should be noted that the ground state energy $3\hbar/2$ is due to the transformation properties of the bundle $\sqrt{\wedge^3 F_B}$ under the one-parameter group of canonical transformations generated by H, cf. the derivation of Eq. (6.24).

The polarization F_B is transverse to the polarization F introduced in Sec. 7.1 which gives rise to the Schrödinger representation. Hence, we can construct the Blattner-Kostant-Sternberg kernel using the technique developed in Sec. 5.1. Substituting $\xi_{\underline{z}}$ for $\underline{\zeta}_1$ and $\underline{\xi}$ given by Eq. (7.3) for $\underline{\zeta}_2$ in Eq. (5.11), we obtain

$$\begin{aligned}&\langle \psi(\underline{q})\lambda_o \otimes \nu_{\tilde{\underline{\xi}}}, \Psi(\underline{z})\lambda_1 \otimes \nu_{\tilde{\xi}_{\underline{z}}}\rangle \\ &= (i\hbar)^{-3/2}2^{-3/4}\exp\{-(2\hbar)^{-1}[|\underline{z}|^2 + i\underline{p}\cdot\underline{q}]\}\overline{\Psi}(\underline{z})\psi(\underline{q}).\end{aligned} \tag{8.36}$$

Substituting this result into Eq. (5.12) we get

$$\mathscr{K}_B(\psi\lambda_o \otimes \nu_{\tilde{\underline{\xi}}}, \psi\lambda_1 \otimes \nu_{\tilde{\xi}_{\underline{z}}}) \tag{8.37}$$
$$= (i/\hbar)^{3/2} 2^{-3/4} \int_{R^6} \exp\{-(2\hbar)^{-1}[|\underline{z}|^2 + i\underline{p}\cdot\underline{q})]\}\overline{\Psi}(\underline{z})\psi(\underline{q})d^3p d^3q.$$

The kernel $\mathscr{K}_B$ induces a linear map $\mathscr{U}_B : \mathscr{A}_B \to \mathscr{A}$ defined by Eq. (5.1) which can be written with the help of Eq. (7.7) as follows:

$$\mathscr{U}_B(\Psi\lambda_1 \otimes \nu_{\tilde{\xi}_{\underline{z}}}) \tag{8.38}$$
$$= (i\hbar)^{-3/2} 2^{-3/4} \left\{\int_{R^3} \exp\{-(2\hbar)^{-1}[|\underline{z}|^2 - i\underline{p}\cdot\underline{q}]\}\Psi(\underline{z})d^3p\right\}\lambda_o \otimes \nu_{\underline{\xi}}.$$

Let U denote the linear operator from the space of holomorphic functions on C^3 square integrable with the weight $\exp(-|\underline{z}|^2/\hbar)$ to the space of square integrable functions on R^3 defined by

$$(U\Psi)(\underline{q}) = (i\hbar)^{-3/2} 2^{-3/4} \int_{R^3} \exp\{-(4\hbar)^{-1}(\underline{p}^2 + \underline{q}^2 - 2i\underline{p}\cdot\underline{q})\}$$
$$\times \Psi(2^{-\frac{1}{2}}(\underline{p} + i\underline{q}))d^3p. \tag{8.39}$$

We can now rewrite Eq. (8.38) in the form

$$\mathscr{U}_B(\Psi(\underline{z})\lambda_1 \otimes \nu_{\tilde{\xi}_{\underline{z}}}) = (U\Psi)(\underline{q})\lambda_o \otimes \nu_{\underline{\xi}}. \tag{8.40}$$

Making use of the identity

$$\Psi(\underline{c}) = (2\pi)^{-3} \int_{R^6} e^{\underline{c}\cdot\overline{\underline{z}}}\, \Psi(\underline{z}) e^{-|\underline{z}|^2} d^3p d^3q \tag{8.41}$$

valid for each holomorphic function Ψ on C^3 and each $\underline{c} \in C^3$, one can verify that U is invertible with inverse given by

$$(U^{-1}\psi)(\underline{z})$$

$$= (-i/\hbar)^{3/2}2^{3/4}\int_{R^3}\exp\{\hbar^{-1}[\tfrac{1}{2}(\underline{z}^2-\underline{v}^2)-i2^{\frac{1}{2}}\underline{z}\cdot\underline{v}]\}\psi(\underline{v})d^3v. \tag{8.42}$$

The mapping U^{-1} corresponds to the unitary transformation of Bargmann from the space of square integrable functions on R^3 to the space of holomorphic functions on C^3 square integrable with respect to the weight $\exp(-|\underline{z}|^2/\hbar)$. Hence, $\mathscr{U}_B : \mathscr{A}_B \to \mathscr{A}$ is unitary and the Bargmann-Fock representation is unitarily related to the Schrödinger representation. Moreover, if $\mathscr{Q}f$ denotes the quantization of a function f in the Schrödinger representation we have

$$(\mathscr{Q}f)\mathscr{U}_B = \mathscr{U}_B(\mathscr{Q}_B f) \tag{8.43}$$

for f equal to q^i, p_i, J_i or the harmonic oscillator energy H.

All the preceding discussion was restricted to the case of three degrees of freedom. Clearly, one can introduce the Bargmann-Fock representation for an arbitrary number of degrees of freedom provided the phase space is an affine symplectic space.

8.2. Harmonic oscillator energy representation

Let us consider a one-dimensional harmonic oscillator with the energy

$$H = \tfrac{1}{2}(p^2 + q^2). \tag{8.44}$$

The Hamiltonian vector field of H with respect to the symplectic form

$$\omega = dp\wedge dq \tag{8.45}$$

on R^2 is given by

$$\xi_H = p \frac{\partial}{\partial q} - q \frac{\partial}{\partial p} . \tag{8.46}$$

To quantize this system in the representation in which the energy is diagonal we require a polarization containing ξ_H. However, the distribution spanned by ξ_H has a singularity at the origin since $\xi_H = 0$ when $p = q = 0$. Since at present we have no theory of quantization for polarizations with singularities, it is necessary to remove the origin and take $X = \mathbf{R}^2 - \{(0,0)\}$ as the phase space of the harmonic oscillator. Since X is open in $\mathbf{R}^2$ we do not introduce special symbols for the restrictions of ω and ξ_H to X.

Let D_H denote the real distribution on X spanned by ξ_H. Since all the integral curves of ξ_H are periodic with period 2π, the integral manifolds of D_H are isomorphic to $\mathbf{T}^1$. The space X/D_H of integral manifolds of D_H is therefore a quotient manifold of X diffeomorphic to $\mathbf{R}^1$. Hence $F_H = D_H{}^{\mathbf{C}}$ is a complete strongly admissible real polarization of (X,ω). Since the integral manifolds of D_H are not simply connected, it follows from the general theory developed in Secs. 4.4 and 4.5 that the Bohr-Sommerfeld variety S is a proper subset of X and that the representation space $\mathcal{H}_H$ will consist of distributional wave functions. In the terminology of Sec. 4.5, we have $X = X_1$ [cf. Eq. (4.39)], and the manifolds $Q \subseteq X/D_H$ intorduced in Eqs. (4.41) and (4.42) will be single points. In fact, the projection $\pi_{D_H}(S)$ of the Bohr-Sommerfeld variety to X/D_H is a sequence $\{Q_n\}$ of isolated points.

The phase space X is not contractible and the additional structure needed for quantization is not unique. In order to obtain results comparable to those derived in the

Bargmann-Fock representation, we choose the trivial metaplectic structure on (X,ω) which is induced by the unique metaplectic structure on $(\mathbf{R}^2,\omega)$. That is, the bundle $\tilde{\mathscr{B}}_\omega\mathscr{T}X$ of metaplectic frames of (X,ω) is the restriction to X of the bundle $\tilde{\mathscr{B}}_\omega\mathscr{T}\mathbf{R}^2$. The bundle $\tilde{\mathscr{B}}_\omega\mathscr{T}X$ induces a metalinear frame bundle $\tilde{\mathscr{B}}F_H$ of F_H in the manner described in Sec. 5.3. We now study the bundle $\tilde{\mathscr{B}}F_H$ in some detail.

The symplectic frame bundle $\mathscr{B}_\omega\mathscr{T}X$ has a trivializing section

$$\beta = \left(\frac{\partial}{\partial p};\frac{\partial}{\partial q}\right). \tag{8.47}$$

Let $\tilde{\beta}$ be a trivializing section of $\tilde{\mathscr{B}}_\omega\mathscr{T}X$ projecting onto β. The sections β and $\tilde{\beta}$ determine isomorphisms $\mathscr{B}_\omega\mathscr{T}X \simeq X \times Sp(1,\mathbf{R})$ and $\tilde{\mathscr{B}}_\omega\mathscr{T}X \simeq X \times Mp(1,\mathbf{R})$ respectively. According to the results of Sec. 5.3, every Lagrangian frame $\underline{w}$ at x can be identified with a mapping $\underline{w}^\#:\mathscr{B}_\omega\mathscr{T}_xX \to B \times GL(1,\mathbf{C})$ satisfying Eq. (5.54), where B is given by Eq. (5.59) and the action of $Sp(1,\mathbf{R})$ on $B \times GL(1,\mathbf{C})$ is defined by Eqs. (5.62), (5.63) and (5.64). Similarly, a metalinear Lagrangian frame at x is given by a mapping $\tilde{\underline{w}}^\#:\tilde{\mathscr{B}}_\omega\mathscr{T}_xX \to B \times ML(1,\mathbf{C})$ satisfying Eq. (5.70). The Hamiltonian vector field ξ_H spans F_H, hence it defines a mapping from $X \times Sp(1,\mathbf{R})$ to $B \times GL(1,\mathbf{C})$ which behaves appropriately under the actions of $Sp(1,\mathbf{R})$ and $GL(1,\mathbf{C})$. Let $\mu: X \to GL(1,\mathbf{C})$ be the mapping obtained by restricting $X \times Sp(1,\mathbf{R}) \to B \times GL(1,\mathbf{C})$ to $X \times \{e\}$ and projecting the result to $GL(1,\mathbf{C})$. Making use of Eqs. (5.47), (5.56), (8.46) and (8.47) we can compute μ explicitly:

$$\mu(x) = -q(x) - ip(x) \in GL(1,\mathbf{C})$$

for each $x \in X$. The bundle $\tilde{\mathscr{B}}F_H$ is trivial if and only if there exists a lift $\tilde{\mu}: X \to ML(1,\mathbf{C})$ of μ. Let us consider a curve $\gamma: [0,2\pi] \to X$ such that $q(\gamma(t)) = -\cos t$ and $p(\gamma(t)) = -\sin t$. Then

$$\mu(\gamma(t)) = e^{it}$$

for each $t \in [0,2\pi]$, and the curve $\mu\circ\gamma: [0,2\pi] \to GL(1,\mathbf{C})$ generates the fundamental group of $GL(1,\mathbf{C})$. Hence the lift of $\mu\circ\gamma$ to the unique curve in $ML(1,\mathbf{C})$ originating from the identity and covering $\mu\circ\gamma$ is not closed. Therefore, the map $\mu: X \to GL(1,\mathbf{C})$ cannot be lifted to a map $\tilde{\mu}: X \to ML(1,\mathbf{C})$ which implies that $\tilde{\mathscr{B}}F_H$ is not trivial.

Consider a covering of X by contractible open sets V^+ and V^-, where

$$V^{\pm} = \{x \in X \mid \pm q(x) > 0 \quad\text{or}\quad p(x) \neq 0\}. \tag{8.48}$$

The restriction of $\tilde{\mathscr{B}}F_H$ to $V^{\pm}$ is trivial and we denote by $\tilde{\xi}_H^{\pm}$ a lift of $\xi_H|V^{\pm}$ to a section of $\tilde{\mathscr{B}}F_H|V^{\pm}$. Moreover, we assume that

$$\tilde{\xi}_H^{+}(x) = \tilde{\xi}_H^{-}(x) \quad\text{when}\quad p(x) > 0 \quad\text{and}\quad q(x) = 0. \tag{8.49}$$

This implies that

$$\tilde{\xi}_H^{+}(x) = \tilde{\xi}_H^{-}(x)\tilde{J} \quad\text{when}\quad p(x) < 0 \quad\text{and}\quad q(x) = 0, \tag{8.50}$$

where $\tilde{J}$ is the element of the kernel of $\rho: ML(1,\mathbf{R}) \to GL(1,\mathbf{R})$ such that

$$\chi(\tilde{J}) = -1. \tag{8.51}$$

The representation space $\mathscr{H}_H$ consists of the space of sections of $L \otimes \sqrt{\wedge}^1 F_H$ covariantly constant along F_H, where L is the trivial complex line bundle over X. We choose a trivializing section λ_o such that

$$\langle\lambda_o, \lambda_o\rangle = 1 \tag{8.52}$$

and

$$\nabla\lambda_o = -i\hbar^{-1} p\,dq \otimes \lambda_o. \tag{8.53}$$

Let $\nu_\pm$ denote the sections of $\sqrt{\wedge^1 F_H}|V^\pm$ defined by

$$\nu_\pm{}^{\#} \circ \tilde{\xi}_H{}^{\pm} = 1. \tag{8.54}$$

Then, every section σ of $L \otimes \sqrt{\wedge^1 F_H}$ can be expressed locally as follows:

$$\sigma|V^\pm = \psi^\pm(\lambda_o|V^\pm) \otimes \nu_\pm, \tag{8.55}$$

where $\psi^\pm$ are complex-valued functions on $V^\pm$. If σ is to be covariantly constant along F_H, the functions $\psi^\pm$ must satisfy the differential equation

$$(p\,\frac{\partial}{\partial q} - q\,\frac{\partial}{\partial p})\psi^\pm - i\hbar^{-1}p^2\,\psi^\pm = 0 \tag{8.56}$$

together with the boundary conditions

$$\psi^+(x) = \psi^-(x) \quad \text{when} \quad p(x) > 0 \quad \text{and} \quad q(x) = 0 \tag{8.57}$$

$$\psi^+(x) = -\psi^-(x) \quad \text{when} \quad p(x) < 0 \quad \text{and} \quad q(x) = 0 \tag{8.58}$$

which follow from Eqs. (8.49) and (8.50), respectively.

Introducing an angular variable ϕ such that

$$p = (p^2 + q^2)^{\frac{1}{2}}\sin\phi, \quad q = (p^2 + q^2)^{\frac{1}{2}}\cos\phi, \tag{8.59}$$

we can write the solutions of (8.56) in the form

$$\psi^+ = a^+\exp[-i\hbar^{-1}(H\phi - \tfrac{1}{2}pq)] \quad \text{for} \quad -\pi < \phi < \pi \tag{8.60}$$

and

$$\psi^- = a^- \exp[-i\hbar^{-1}(H\phi - \tfrac{1}{2}pq)] \quad \text{for} \quad 0 < \phi < 2\pi, \tag{8.61}$$

where $a^{\pm}$ are amplitudes depending only on H. Taking into account the boundary conditions (8.57) and (8.58) we have

$$a^+ \exp[i\hbar^{-1}H\pi/2] = a^- \exp[-i\hbar^{-1}H\pi/2] \tag{8.62}$$

and

$$a^+ \exp[i\hbar^{-1}H(-\pi/2)] = -a^- \exp[-i\hbar^{-1}H3\pi/2]. \tag{8.63}$$

Hence

$$a^+ = a^- \tag{8.64}$$

and

$$a^+(H) \exp[-i\hbar^{-1}H2\pi] = -a^+(H) \tag{8.65}$$

which implies that $a^+(H(x)) = 0$ unless $H(x) = (n + \tfrac{1}{2})\hbar$ for some integer n. Thus, the supports of the covariantly constant sections of $L \otimes \sqrt{\wedge^1 F_H}$ are contained in the Bohr-Sommerfeld variety

$$S = \{x \in X | H(x) = (n + \tfrac{1}{2})\hbar, \; n \in Z\}. \tag{8.66}$$

This result could be obtained directly from Eq. (4.38) by substituting $\tfrac{1}{2}$ for d_γ on account of non-triviality of $\tilde{\mathscr{B}}F_H$.

The Bohr-Sommerfeld variety S consists of a sequence of concentric circles $H^{-1}[(n + \tfrac{1}{2})\hbar]$. Hence, the scalar product on $\mathscr{H}_H$ is given by Eq. (4.46) with $k = 1$. For each non-negative integer n, we denote by σ_n a covariantly constant section of $L \otimes \sqrt{\wedge^1 F_H} | H^{-1}[(n + \tfrac{1}{2})\hbar]$ such that

$$(\sigma_n | \sigma_n) = 1. \tag{8.67}$$

For each $x \in H^{-1}[(n + \tfrac{1}{2})\hbar] \cap V^{\pm}$ we can write

$$\sigma_n(x) = c_n \exp\{-i(n+\tfrac{1}{2})[\phi_\pm(x) - \tfrac{1}{2}\sin 2\phi_\pm(x)]\}\lambda_0(x) \otimes \nu_\pm(x), \tag{8.68}$$

where $\phi_+(x)$ and $\phi_-(x)$ are the values of the variable ϕ satisfying Eq. (8.59) corresponding to the point $x \in X$ and contained in the intervals $-\pi < \phi_+(x) < \pi$ and $0 < \phi_-(x) < 2\pi$.

The value of the normalization constant c_n can be obtained from Eqs. (4.46) to (4.51). Since the integral curves of ξ_H are periodic with period 2π, we have $\kappa(\xi_H) = (2\pi)^{-1}$ so that $\kappa(\xi) = 1$ for $\xi = 2\pi\xi_H$. Let Q_n be the projection of $H^{-1}[(n + \frac{1}{2})\hbar]$ to X/D_H and x any point in $H^{-1}[(n + \frac{1}{2})\hbar]$. Eqs. (4.51) and (8.68) yield

$$\langle\sigma_n, \sigma_n\rangle_{Q_n} = |c_n|^2 \, \nu_\pm^{\#}(\tilde{\xi}(x))\overline{\nu_\pm^{\#}(\tilde{\xi}(x))}, \tag{8.69}$$

where $\tilde{\xi}(x)$ is a lift of $\xi(x)$ to $\tilde{\mathscr{B}}F_H$. As $\xi(x) = 2\pi\xi_H(x)$, we obtain from Eqs. (4.12), (8.54) and (8.69)

$$\langle\sigma_n, \sigma_n\rangle_{Q_n} = |c_n|^2 \, (2\pi)^{-1}. \tag{8.70}$$

Since Q_n is a point in X/D_H, there is no integration involved in Eq. (4.46) and we have

$$(\sigma_n|\sigma_n) = \langle\sigma_n, \sigma_n\rangle_{Q_n}. \tag{8.71}$$

Taking into account Eqs. (8.67) and (8.70), we see that we can choose

$$c_n = (2\pi)^{\frac{1}{2}}. \tag{8.72}$$

The sequence (σ_n) of sections in $\mathscr{H}_H$ forms an orthonormal basis of $\mathscr{H}_H$; the orthogonality of σ_n and σ_m with $m \neq n$ is obvious since $(\text{support } \sigma_n) \cap (\text{support } \sigma_m) = \phi$. Hence, each $\sigma \in \mathscr{H}_H$ can be written

$$\sigma = \sum_n a_n \sigma_n, \tag{8.73}$$

where the a_n are complex numbers, and the scalar product on $\mathscr{A}_H$ is given by

$$(\sum_n a_n \sigma_n | \sum_m b_m \sigma_m) = \sum_n a_n \overline{b}_n. \tag{8.74}$$

Thus, the basis (σ_n) establishes an isomorphism between $\mathscr{A}_H$ and the Hilbert space of square summable sequences.

The quantized Hamiltonian $\mathscr{Q}_H H$ in the *energy representation* $\mathscr{A}_H$ acts by multiplication by H, i.e., for each $\sigma \in \mathscr{A}_H$,

$$\mathscr{Q}_H H[\sigma] = H\sigma \tag{8.75}$$

[cf. Eq. (6.26)]. Hence, the vector σ_n is an eigenvector of $\mathscr{Q}_H H$ with eigenvalue $(n + \frac{1}{2})\hbar$, and the spectrum of $\mathscr{Q}_H H$ is given by the image under H of the Bohr-Sommerfeld variety S defined by Eq. (8.66). The quantization of the position and the momentum variables, which is straightforward in the Schrödinger and the Bargmann-Fock representations, cannot be obtained at present in the energy representation for the following reason. Let $\phi_p^{\ t}$ denote the one-parameter local group of local canonical transformations of (X,ω) generated by p; it is a local group since the Hamiltonian vector field of p is not complete in X. The transformations $\phi_p^{\ t}$ correspond to translations along the q-axis. Since the integral manifolds of D_H are the circles $p^2 + q^2 = \text{const}$, the distributions F_H and $\mathscr{T}\phi_p^{\ t}(\overline{F}_H)$ do not intersect along a distribution. Hence, the polarizations F_H and $\mathscr{T}\phi_p^{\ t}(\overline{F}_H)$ do not satisfy the conditions necessary for the construction of the Blattner-Kostant-Sternberg kernels. It may be that this is only a technical difficulty and in the future one will be able to generalize the construction of the Blattner-Kostant-Sternberg kernels to the case of polarizations with singular intersections. However, if this generalization $\mathscr{K}$ is local in X, i.e., if $\mathscr{K}(\sigma_1,\sigma_2)$ is an integral of a concomitant $<\sigma_1,\sigma_2>$ depending locally on the sections σ_1 and σ_2 as in Eq. (5.11), then the resulting operator $\mathscr{Q}_H p$ would commute with $\mathscr{Q}_H H$ according to the argument in Sec. 6.4. Thus, either the generalization of the Blattner-Kostant-Sternberg kernel used in the quantization of the momentum in the energy representation is non-local or the operators $\mathscr{Q}_H p$ and $\mathscr{Q}_H H$ commute, in which case the energy representation is inequivalent to the Schrödinger representation. The same argument is valid for any dynamical variable f on X.

Let F_B be the polarization of (X,ω) spanned by the Hamiltonian vector field ξ_z of the complex coordinate

$$z = 2^{-\frac{1}{2}}(p + iq). \tag{8.76}$$

The polarization F_B gives rise to the Bargmann-Fock representation for the harmonic oscillator in the manner described in Sec. 8.1; the only difference being that now we have one degree of freedom instead of three. The polarizations F_H and F_B satisfy the transversality condition (5.12) so that we can use Eqs. (5.11) and (5.13) to define the Blattner-Kostant-Sternberg kernel $\mathscr{K}_{BH}: \mathscr{H}_B \times \mathscr{H}_H \to \mathbf{C}$. Since the normalized eigenvectors of $\mathscr{Q}_B H$ form an orthonormal basis of $\mathscr{H}_B$, we can describe $\mathscr{K}_{BH}$ by its values K_{mn} on the energy eigenvectors in $\mathscr{H}_B$ and $\mathscr{H}_H$ with the eigenvalues $(m + \frac{1}{2})\hbar$ and $(n + \frac{1}{2})\hbar$ respectively:

$$K_{mn} := \mathscr{K}_{BH}(\sigma_m', \sigma_n), \tag{8.77}$$

where σ_n is given by Eq. (8.68) and

$$\sigma_m{}' = \Psi_m(z)\lambda_1 \otimes \nu_{\tilde{\xi}_z} . \tag{8.78}$$

Here, λ_1 is defined by Eq. (8.11) and Ψ_1 satisfies the equation

$$z \frac{\partial \Psi_m}{\partial z} = m\Psi_m \tag{8.79}$$

which follows from Eq. (8.35) if one takes into account the fact that we have one degree of freedom only. Thus

$$\Psi_m = a_m z^m , \tag{8.80}$$

where the normalization constant a_m is determined by the requirement

$$(\sigma_m{}'|\sigma_m{}') = 1 \tag{8.81}$$

which, in view of Eq. (8.19), is equivalent to

$$|a_m|^2 \int_{\mathbf{R}^2} |z|^{2m} \exp(-|z|^2/\hbar)\,dp\,dq = 1. \tag{8.82}$$

Evaluating the integral, we find that we can choose

$$a_m = (2\pi\ m!\ \hbar^{m+1})^{-\frac{1}{2}}. \tag{8.83}$$

Substituting Eqs. (8.68) and (8.78) into Eq. (5.11) and taking into account Eq. (8.11), we obtain

$$\langle \sigma_m{}', \sigma_n \rangle = b_{mn} \exp[i\phi(n-m)] \tag{8.84}$$

where the b_{mn} are constants different from zero. The matrix coefficient K_{mn} defined by Eq. (8.77) can be evaluated with the help of Eq. (5.13) where the integration is taken over $H^{-1}[(n + \frac{1}{2})\hbar]$ since in this case $Q_1 = X$ and Q_2 is the single point in X/D_H corresponding to the energy $(n + \frac{1}{2})\hbar$. The density $\delta_{Q_1Q_2}$ in Eq. (5.13) corresponds to integration with respect to $\phi \in [0,2\pi]$. Thus K_{mn} is proportional to $\int_0^{2\pi} \exp[i\phi(n-m)]d\phi$ which implies that

$$K_{mn} = 0 \quad \text{for} \quad m \neq n. \tag{8.85}$$

Hence, the linear mapping $\mathscr{U}_{BH}: \mathscr{A}_H \to \mathscr{A}_B$ defined according to Eq. (5.1) by $\mathscr{K}_{BH}$ satisfies

$$\mathscr{U}_{BH}(\mathscr{Q}_H H) = (\mathscr{Q}_B H)\mathscr{U}_{BH}. \tag{8.86}$$

The diagonal elements K_{mm} are non-zero which implies that $\mathscr{U}_{BH}$ is a vector space isomorphism. On the other hand, $|K_{mm}| \neq 1$ so that $\mathscr{U}_{BH}$ does not preserve the scalar product and, therefore, is not unitary.

9. TIME-DEPENDENT SCHRÖDINGER EQUATION

Time-dependent Hamiltonian dynamics can be formulated in evolution space independently of any inertial frame. The principle of geometric quantization extended to evolution space yields an intrinsic quantum theory equivalent to that based on the time-dependent Schrödinger equation. We restrict our attention to the quantum mechanics of a single particle with a time-dependent potential.

Following the terminology of Sec. 2.4, we denote by Z the evolution space of the particle. That is,

$$Z = \bigcup_{t \in T} X_t , \tag{9.1}$$

where T represents absolute time and X_t is the phase space at time t equipped with the Lagrange bracket ω_t. Since we are dealing with a single particle without constraints, each X_t is the cotangent bundle space of a 3-dimensional Euclidean space Y_t representing the physical space at time t,

$$X_t = \mathcal{T}^* Y_t , \tag{9.2}$$

and

$$\omega_t = d\theta_{Y_t} \tag{9.3}$$

where θ_{Y_t} is the canonical 1-form on $\mathscr{T}^*Y_t$ defined by Eq. (2.12).

In each of the phase spaces (X_t, ω_t) we introduce the structure leading to the Schrödinger representation, cf. Sec. 7.1. Thus, we have a trivial bundle $\tilde{\mathscr{B}}_{\omega_t}\mathscr{T}X_t$ of metaplectic frames for (X_t, ω_t) and a polarization F_t spanned over $\mathbb{C}$ by the vectors tangent to the cotangent bundle projection $X_t = \mathscr{T}^*Y_t \to Y_t$. The bundle $\tilde{\mathscr{B}}_{\omega_t}\mathscr{T}X_t$ induces a trivial complex line bundle $\sqrt{\wedge^3 F_t}$ over X_t. Moreover, we have a complex line bundle L_t over (X_t, ω_t) with a connection ∇ satisfying the prequantization condition (3.20) and also a trivializing section $\lambda_t\colon X_t \to L_t$ such that

$$\nabla\lambda_t = -i\hbar^{-1}\theta_{Y_t} \otimes \lambda_t \tag{9.4}$$

and

$$\langle\lambda_t, \lambda_t\rangle = 1. \tag{9.5}$$

The representation space defined by this structure is denoted $\mathscr{H}_t$.

The collection of fibre bundles introduced above gives rise to fibre bundles $\tilde{\mathscr{B}}_{\omega}\mathscr{T}Z$, F, $\sqrt{\wedge^3 F}$ and L over Z such that, for each $t \in T$,

$$\tilde{\mathscr{B}}_{\omega}\mathscr{T}Z \,|\, X_t = \tilde{\mathscr{B}}_{\omega_t}\mathscr{T}X_t \tag{9.6}$$

$$F \,|\, X_t = F_t \tag{9.7}$$

$$\sqrt{\wedge^3 F} \,|\, X_t = \sqrt{\wedge^3 F_t} \tag{9.8}$$

and

$$L \,|\, X_t = L_t. \tag{9.9}$$

The sections $\{\lambda_t\colon X_t \to L_t \mid t \in T\}$ induce a section $\lambda\colon X \to L$ such that

$$\lambda | X_t = \lambda_t \tag{9.10}$$

for each $t \in T$. We assume that all these bundles are differentiable and that λ is a smooth section of L. The connections in the bundles L_t give rise to a partial connection in L. We extend it to a connection ∇ by requiring that

$$\nabla\lambda = -i\hbar^{-1}\Theta \otimes \lambda, \tag{9.11}$$

where Θ is a one-form on Z such that

$$\Theta | X_t = \theta_{Y_t} \tag{9.12}$$

for each $t \in T$, and

$$\Omega = d\Theta \tag{9.13}$$

is the 2-form determining the classical dynamics of the particle in the manner described in Sec. 2.4. The classical motions are the orbits of the one-parameter group ϕ^s of diffeomorphisms of Z generated by the vector field ζ on Z satisfying

$$\zeta \rfloor \Omega = 0 \tag{9.14}$$

and normalized by Eq. (2.41).

The *quantum evolution space* $\mathscr{E}$ is the union of the representation spaces $\mathscr{H}_t$,

$$\mathscr{E} = \bigcup_{t \in T} \mathscr{H}_t. \tag{9.15}$$

It is a bundle of Hilbert spaces over T. The quantum dynamics of a single particle is given by a trivialization of this bundle: sections of $\mathscr{E} \to T$ correspond to sections of $L \otimes \sqrt{\wedge^3 F}$ which are covariantly constant along F. In order to describe the quantum dynamics induced by the classical dynamics given by ϕ^s, we shall associate to ϕ^s a one-parameter family of vector bundle maps $\Phi_s: \mathscr{S} \to \mathscr{S}$. The quantum dynamics is then determined by Φ_s as follows: A section σ of $\mathscr{S}$ describes a dynamically admissible history of the quantum system under consideration if and only if

$$\frac{d}{ds}(\Phi_s \sigma)\Big|_{s=0} = 0. \tag{9.16}$$

Let $\mathscr{A}$ denote the space of solutions of Eq. (9.16). For each $t \in T$, we have the restriction mapping $\mathscr{U}_t: \mathscr{A} \to \mathscr{A}_t$ defined by

$$\mathscr{U}_t \sigma = \sigma | X_t. \tag{9.17}$$

We shall see later that Eq. (9.16) corresponds to the time-dependent Schrödinger equation. Hence, the space $\mathscr{A}$ admits a Hilbert space structure such that all the mappings $\mathscr{U}_t: \mathscr{A} \to \mathscr{A}_t$ are unitary.

Let f be a function on Z such that, for each $t \in T$, the restriction f_t of f to X_t can be quantized in the Schrödinger representation giving rise to an operator $\mathscr{Q}f_t$ on $\mathscr{A}_t$. We denote by $\mathscr{Q}_t f$ the one-parameter family of operators on $\mathscr{A}$ defined by

$$\mathscr{Q}_t f = \mathscr{U}_t^{-1}(\mathscr{Q}f_t)\mathscr{U}_t. \tag{9.18}$$

The operators $\mathscr{Q}_t f$ do not depend on $t \in T$ if, for each t and $t' \in T$,

$$(\mathscr{Q}f_t)\mathscr{U}_t\mathscr{U}_{t'}^{-1} = \mathscr{U}_t\mathscr{U}_{t'}^{-1}(\mathscr{Q}f_{t'}). \tag{9.19}$$

In this case there exists an operator $\mathscr{Q}f$ such that

$$\mathscr{Q}_t f = \mathscr{Q}f \tag{9.20}$$

for all $t \in T$.

It remains to construct the vector bundle maps $\Phi_s: \mathscr{S} \to \mathscr{S}$ which determine the quantum dynamics. The one-parameter group ϕ^s generated by the vector field ζ satisfying Eqs. (9.14) and (2.41) preserves Ω and it induces a one-parameter group of translations of T denoted by $t \mapsto t + s$. Hence, we can lift the action of ϕ^s to the bundle

$$\tilde{\mathscr{P}}_{\omega}\mathscr{T}Z = \bigcup_{t\in T} \tilde{\mathscr{P}}_{\omega_t}\mathscr{T}X_t, \tag{9.21}$$

where $\tilde{\mathscr{P}}_{\omega_t}\mathscr{T}X_t$ is the bundle of positive metalinear Lagrangian frames for (X_t,ω_t). We denote by $\tilde{\phi}^{\#s}$ the one-parameter group of automorphisms of $\tilde{\mathscr{P}}_{\omega}\mathscr{T}Z$ induced by ϕ^s. The distribution F is not invariant under the action of ϕ^s and we denote by $\mathscr{T}\phi^s(F)$ the image of F under ϕ^s. For each $t \in T$, the distribution $\mathscr{T}\phi^s(F)|X_t$ is a polarization of (X_t,ω_t) and

$$\mathscr{T}\phi^s(F)|X_t = \mathscr{T}\phi^s(F_{t-s}). \tag{9.22}$$

Let, for each $t \in T$, $\tilde{\mathscr{B}}\mathscr{T}\phi^s(F_{t-s})$ be the bundle of metalinear frames for $\mathscr{T}\phi^s(F_{t-s})$ induced by the metaplectic frame bundle $\tilde{\mathscr{B}}_{\omega_t}\mathscr{T}X_t$, and $\sqrt{\Lambda^3\mathscr{T}\phi^s(F_{t-s})}$ the associated bundle corresponding to the character χ of $ML(3,\mathbf{C})$. We denote by $\sqrt{\Lambda^3\mathscr{T}\phi^s(F)}$ the complex line bundle over Z such that, for each $t \in T$,

$$\sqrt{\Lambda^3\mathscr{T}\phi^s(F)}|X_t = \sqrt{\Lambda^3\mathscr{T}\phi^s(F_{t-s})}. \tag{9.23}$$

For each section ν of $\sqrt{\Lambda^3 F}$, we define the section $\phi^s\nu$ of $\sqrt{\Lambda^3\, \phi^s(F)}$ by

$$(\phi^s\nu)^{\#}(\underline{\tilde{w}}) = \nu^{\#}(\tilde{\phi}^{\#-s}(\underline{\tilde{w}})) \tag{9.24}$$

for each $\underline{\tilde{w}} \in \tilde{\mathscr{B}}\mathscr{T}\phi^s(F) = \bigcup_{t\in T} \tilde{\mathscr{B}}\mathscr{T}\phi^s(F_{t-s})$.

In a similar manner we can define the action of ϕ^s on sections of L. Let $\zeta^{\#}$ denote the horizontal lift of ζ to the principal $\mathbf{C}^{\times}$ bundle $L^{\times}$ associated to L

[cf. Eq. (3.13)], and $\phi^{\#s}$ the one-parameter group of diffeomorphisms of $L^\times$ generated by $\zeta^\#$. For each section λ of L, the section $\phi^s\lambda$ is defined by

$$(\phi^s\lambda)^\# = \lambda^\# \circ \phi^{\#-s}. \tag{9.25}$$

Let $\mathscr{H}_t^s$ denote the representation space defined by the polarization $\mathscr{T}\phi^s(F)|X_t$ of (X_t,ω_t). Each section σ_{t-s} in $\mathscr{H}_{t-s}$ can be written

$$\sigma_{t-s} = \lambda_{t-s} \otimes \nu_{t-s}, \tag{9.26}$$

where λ_{t-s} is the trivializing section of L_{t-s} satisfying Eqs. (9.4) and (9.5) and ν_{t-s} is a section of $\sqrt{\wedge^3 F_{t-s}}$ covariantly constant along F_{t-s}. The action of ϕ^s on sections of L and $\sqrt{\wedge^3 F}$ yields a unitary transformation $\Phi_t^s\colon \mathscr{H}_{t-s} \to \mathscr{H}_t^s$ defined by

$$\Phi_t^s(\sigma_{t-s}) = \phi^s\lambda_{t-s} \otimes \phi^s\nu_{t-s}. \tag{9.27}$$

We assume that, for sufficiently small $s > 0$, the polarizations $\mathscr{T}\phi^s(F_{t-s})$ and F_t are transverse so that one can construct the Blattner-Kostant-Sternberg kernel $\mathscr{K}_{t,s}\colon \mathscr{H}_t \times \mathscr{H}_t^s \to \mathbb{C}$ as in Sec. 5.1. Let $\mathscr{U}_{t,s}\colon \mathscr{H}_t^s \to \mathscr{H}_t$ be the linear map induced by $\mathscr{K}_{t,s}$ such that, for each $\sigma_t \in \mathscr{H}_t$ and each $\sigma_t^s \in \mathscr{H}_t^s$,

$$\mathscr{K}_{t,s}(\sigma_t,\sigma_t^s) = (\sigma_t|\mathscr{U}_{t,s}\sigma_t^s). \tag{9.28}$$

Then $\mathscr{U}_{t,s}\circ\Phi_t^s$ is a linear map from $\mathscr{H}_{t-s}$ to $\mathscr{H}_t$. The collection $\{\mathscr{U}_{t,s}\circ\Phi_t^s\colon \mathscr{H}_{t-s} \to \mathscr{H}_t \mid t \in T\}$ of linear maps induces a vector bundle map $\Phi_s\colon \mathscr{E} \to \mathscr{E}$ such that, for each $t \in T$,

$$\Phi_s \mid \mathscr{H}_{t-s} = \mathscr{U}_{t,s}\circ\Phi_t^s. \tag{9.29}$$

Let $Y \approx E \times T$ be the product structure of the space-time Y induced by the choice of some inertial frame; it gives rise to a product structure on Z with the projection $\upsilon: Z \to \mathcal{T}^*E$ and $\tau: Z \to T$. The choice of an initial time yields an affine isomorphism $t: T \to \mathbb{R}$ such that the 1-form dt, corresponding to the choice of the time scale, coincides with the differential of the function t. We denote by $\underline{q} = (q_1, q_2, q_3)$ the pull-back to Z of cartesian coordinates on E and by $\underline{p} = (p_1, p_2, p_3)$ the conjugate momentum functions. We shall also use the letter t to denote the pull-back of $t: T \to \mathbb{R}$ to Z. The functions $\underline{p}$, $\underline{q}$ and t form a coordinate system on Z. The distribution F is spanned by the linear frame field

$$\underline{\xi} = (\xi^1, \xi^2, \xi^3), \tag{9.30}$$

where

$$\xi^i = \frac{\partial}{\partial p_i} \tag{9.31}$$

for $i = 1,2,3$. We denote by $\underline{\tilde{\xi}}$ the metalinear frame field for F projecting onto $\underline{\xi}$ and by $\nu_{\underline{\tilde{\xi}}}$ the trivializing section of $\sqrt{\wedge}^3 F$ such that

$$\nu_{\underline{\tilde{\xi}}}^{\#} \circ \underline{\tilde{\xi}} = 1. \tag{9.32}$$

A section σ of $L \otimes \sqrt{\wedge}^3 F$ covariantly constant along F can be written in the form

$$\sigma = \psi(\underline{q}, t)\lambda \otimes \nu_{\underline{\tilde{\xi}}}\ , \tag{9.33}$$

where λ is the trivializing section of L satisfying Eq. (9.11). Let ψ_s be the function on $\mathbb{R}^4$ such that

$$\Phi_s \sigma = \psi_s(\underline{q}, t)\lambda \otimes \nu_{\underline{\tilde{\xi}}}. \tag{9.34}$$

Then, using reasoning analogous to that leading to Eq. (6.49), we obtain

$$\psi_s(\underset{\sim}{q},t) = (i\hbar)^{-3/2}\int_{\mathbf{R}^3}\Big\{[\det \Omega(\xi^j,\phi^s\xi^k)]^{\frac{1}{2}} \times \exp\Big[i\hbar^{-1}\int_0^s \Theta(\zeta)\circ\phi^{-u}du\Big]\psi(\underset{\sim}{q}\circ\phi^{-s},t-s)\Big\}d^3p. \tag{9.35}$$

If the form Θ is given by

$$\Theta = \sum_j p_j dq^j - [(\underset{\sim}{p}^2/2m) + V(\underset{\sim}{q},t)]dt, \tag{9.36}$$

we can use the arguments leading to Eq. (7.38) to obtain

$$\lim_{s\to 0+}\frac{d}{ds}\psi_s(\underset{\sim}{q},t) = -i\hbar^{-1}V(\underset{\sim}{q},t)\psi(\underset{\sim}{q},t)+(i\hbar/2m)\Delta\psi(\underset{\sim}{q},t) - \frac{\partial}{\partial t}\psi(\underset{\sim}{q},t). \tag{9.37}$$

Substituting Eqs. (9.33), (9.34) and (9.37) into Eq. (9.16), we find

$$i\hbar\frac{\partial}{\partial t}\psi(\underset{\sim}{q},t) = [-(\hbar^2/2m)\Delta + V(\underset{\sim}{q},t)]\psi(\underset{\sim}{q},t). \tag{9.38}$$

Thus, a section $\psi(\underset{\sim}{q},t)\lambda \otimes \nu_{\underset{\sim}{\xi}}$ of $\mathscr{L}$ describes a dynamically admissible history if and only if the function $\psi(\underset{\sim}{q},t)$ satisfies the time-dependent Schrödinger equation.

10. RELATIVISTIC DYNAMICS IN AN ELECTROMAGNETIC FIELD

10.1. Relativistic quantum dynamics

The relativistic dynamics of a particle with charge e in an external electromagnetic field f can be described in terms of the phase space $(\mathcal{T}^*Y, \omega_e)$, where Y is the space-time manifold, $\pi: \mathcal{T}^*Y \to Y$ is the cotangent bundle projection, and

$$\omega_e = d\theta_Y + e\pi^*f \tag{10.1}$$

[cf. Sec. 2.3]. Assuming that Y is orientable, and following the reasoning of Sec. 7.2 leading to a metaplectic structure on $(\mathcal{T}^*Y, d\theta_Y)$, we obtain a metaplectic structure on $(\mathcal{T}^*Y, \omega_e)$. The vertical distribution D on $\mathcal{T}^*Y$ tangent to the fibres of π is Lagrangian with respect to the symplectic form ω_e, so that $F = D^{\mathbf{C}}$ is a polarization of $(\mathcal{T}^*Y, \omega_e)$. The metalinear structure of F induced by the metaplectic structure on $(\mathcal{T}^*Y, \omega_e)$ is isomorphic to that induced by the metaplectic structure on $(\mathcal{T}^*Y, d\theta_Y)$. Hence, we can apply the results of Sec. 7.2. We denote by $\tilde{\mathcal{B}}F$ the metalinear frame bundle of F induced by the metaplectic structure and by $\sqrt{\wedge^4 F}$ the associated line bundle corresponding to the character χ of $ML(4, \mathbf{C})$.

For each local chart $(V_\alpha, \check{q}_\alpha^{\ 1}, \ldots, \check{q}_\alpha^{\ 4})$ on Y the restriction of F to $\pi^{-1}(V_\alpha)$ is spanned by the linear frame field

$$\xi_{\underline{q}_\alpha} = (\xi_{q_\alpha^{\ 1}}, \ldots, \xi_{q_\alpha^{\ 4}}). \tag{10.2}$$

Note that the Hamiltonian vector field of a function $q = \check{q} \circ \pi$, where $\check{q}$ is a function on Y, does not depend on the value of e in Eq. (10.1). We have a family ν_α of local sections of $\sqrt{\wedge^4 F}$ such that

$$\nu_\alpha^{\ \#} \circ \tilde{\xi}_{\underline{q}_\alpha} = 1, \tag{10.3}$$

where $\tilde{\xi}_{\underline{q}_\alpha}$ is a lift of $\xi_{\underline{q}_\alpha}$ to a metalinear frame field. There exists a global trivializing section ν_g of $\sqrt{\wedge^4 F}$ satisfying Eq. (7.70), where g is the space-time metric.

Let L_e be a prequantization line bundle corresponding to the symplectic form ω_e. The prequantization condition (3.20) takes the form

$$d\alpha_e = -\hbar^{-1}\pi^*\omega_e, \tag{10.4}$$

where α_e is the connection form on L_e and $\pi^*\omega_e$ denotes the pull-back of ω_e to $L_e^{\times}$. It is satisfied if and only if the de Rham cohomology class $[\hbar^{-1}\omega_e]$ defined by the symplectic form $\hbar^{-1}\omega_e$ is integral. Eq. (10.1) yields

$$[\hbar^{-1}\omega_e] = \pi^*[\hbar^{-1}ef], \tag{10.5}$$

where the right hand side denotes the pull-back of the cohomology class $[\hbar^{-1}ef]$ on Y to $\mathscr{T}^*Y$. Hence, the quantization condition (10.4) is satisfied if and only if $[\hbar^{-1}ef]$ is an integral class in $H^2(Y,\mathbb{R})$. This implies that, for each compact oriented 2-surface Σ in Y with empty boundary

$$\int_\Sigma h^{-1} ef = k_\Sigma \in Z. \tag{10.6}$$

In the presence of magnetic charges the form f representing the electromagnetic field strength ceases to be closed. Therefore, the preceding argument applies only if Y represents the part of space-time free of magnetic charges. Since

$$\int_\Sigma f = 4\pi m_\Sigma, \tag{10.7}$$

where m_Σ is the total magnetic charge surrounded by Σ, we have

$$em_\Sigma = \tfrac{1}{2} k_\Sigma \hbar \tag{10.8}$$

for some integer k_Σ. Eq. (10.8) has to be satisfied for each closed surface Σ without boundary. This is possible only if m_Σ is an integral multiple of some magnetic charge m satisfying the *Dirac condition*

$$2emh^{-1} \in Z. \tag{10.9}$$

Since the trivializing section ν_g of $\sqrt{\wedge^4 F}$ is covariantly constant along F, every section of $L \otimes \sqrt{\wedge^4 F}$ covariantly constant along F is of the form $\lambda \otimes \nu_g$, where λ is a covariantly constant section of L_e. There is a relation $\sim$ in L_e given by $\ell_1 \sim \ell_2$ if and only if there exists $y \in Y$ such that ℓ_1 can be joined to ℓ_2 by a horizontal curve in $L_e | \mathscr{T}_y{}^*Y$. Since, for each $y \in Y$, the connection in $L_e | \mathscr{T}_y{}^*Y$ is flat with trivial holonomy group, the relation $\sim$ is an equivalence relation. The space of equivalence classes, denoted $L_e^\sim$, has the structure of a complex line bundle over Y. The canonical projection $\pi^\sim : L_e \to L_e^\sim$ is an isomorphism on each fibre and the diagram

$$\begin{array}{ccc} L_e & \xrightarrow{\pi^\sim} & L_e^\sim \\ \downarrow & & \downarrow \\ \mathscr{T}^*Y & \xrightarrow{\pi} & Y \end{array}, \tag{10.10}$$

in which vertical arrows denote the line bundle projections, commutes. There is a bijection between the space of sections $\lambda^\sim$ of $L_e^\sim$ and the space of covariantly constant sections λ of L_e. If λ is a section of L_e covariantly constant along F, then $\lambda^\sim : Y \to L_e^\sim$ is given by

$$\lambda^\sim(y) = \pi^\sim \circ \lambda(x) \tag{10.11}$$

for each $y \in Y$ and any $x \in \mathscr{T}_y^*Y$. There is an induced Hermitian form $\langle,\rangle^\sim$ on $L_e^\sim$ such that

$$\langle\lambda_1,\lambda_2\rangle = \langle\lambda_1^\sim,\lambda_2^\sim\rangle^\sim \circ \pi \tag{10.12}$$

for each pair (λ_1,λ_2) of covariantly constant sections of L. The representation space $\mathscr{H}_e$ consists of covariantly constant sections $\lambda \otimes \nu_g$ of $L_e \otimes \sqrt{\wedge^4 F}$ with the scalar product

$$(\lambda_1 \otimes \nu_g | \lambda_2 \otimes \nu_g) = \int_Y \langle\lambda_1^\sim,\lambda_2^\sim\rangle^\sim |\det g|^{\frac{1}{2}}. \tag{10.13}$$

If the form ef is not exact, the line bundle L_e does not admit a global nonvanishing section. On each contractible coordinate neighborhood V_α in Y there exists a 1-form $\mathscr{A}_\alpha$ such that

$$f|V_\alpha = d\mathscr{A}_\alpha . \tag{10.14}$$

The form $\mathscr{A}_\alpha$ is a local potential for the electromagnetic field strength f. Let λ_α be a local section of $L_e|\pi^{-1}(V_\alpha)$ such that

$$\nabla\lambda_\alpha = i\hbar^{-1}(\theta_Y + e\pi^*\mathscr{A}_\alpha) \otimes \lambda_\alpha, \tag{10.15}$$

where $\pi^*\mathscr{A}_\alpha$ denotes the pull-back of $\mathscr{A}_\alpha$ to $\pi^{-1}(V_\alpha)$, and

$$\langle\lambda_\alpha, \lambda_\alpha\rangle = 1. \tag{10.16}$$

The section λ_α is covariantly constant along $F|\pi^{-1}(V_\alpha)$, and each section $\lambda \otimes \nu_g \in \mathscr{H}_e$ can be expressed locally as

$$\lambda \otimes \nu_g|\pi^{-1}(V_\alpha) = \Psi_\alpha \otimes \lambda_\alpha \otimes \nu_g, \tag{10.17}$$

where Ψ_α is a complex function on V_α. If λ_1 and λ_2 have supports in $\pi^{-1}(V_\alpha)$ then

$$(\lambda_1 \otimes \nu_g|\lambda_2 \otimes \nu_g) = \int_{V_\alpha} \Psi_{1\alpha}\overline{\Psi}_{2\alpha}\, |\det g|^{\frac{1}{2}}. \tag{10.18}$$

If $V_\alpha \cap V_\beta \neq \phi$ there exists a function $c_{\alpha\beta}\colon V_\alpha \cap V_\beta \to \mathbb{C}^\times$ such that

$$\lambda_\alpha(x) = c_{\alpha\beta}(\pi(x))\lambda_\beta(x) \tag{10.19}$$

for each $x \in \pi^{-1}(V_\alpha \cap V_\beta)$. Eq. (10.16) implies that $|c_{\alpha\beta}| = 1$ and, assuming that $V_\alpha \cap V_\beta$ is contractible, we can find a real-valued function $\Lambda_{\alpha\beta}$ on $V_\alpha \cap V_\beta$ such that

$$c_{\alpha\beta} = \exp(-ie\hbar^{-1}\Lambda_{\alpha\beta}). \tag{10.20}$$

It follows from Eq. (10.15) that

$$\mathscr{A}_\alpha - \mathscr{A}_\beta = d\Lambda_{\alpha\beta} \quad \text{on} \quad V_\alpha \cap V_\beta. \tag{10.21}$$

Since the sections λ_α are covariantly constant along F, they induce local sections $\lambda_\alpha^\sim\colon V_\alpha \to L_e^\sim$. For each $y \in V_\alpha \cap V_\beta$, Eqs. (10.11) and (10.18) yield

$$\lambda_\alpha^\sim(y) = c_{\alpha\beta}(y)\lambda_\beta^\sim(y) \tag{10.22}$$

so that the $c_{\alpha\beta}: V_\alpha \cap V_\beta \to \mathbf{C}^\times$ are transition functions for the line bundle $L_e^\sim$. Eqs. (10.20), (10.21) and (10.22) imply that we can define a connection $\nabla^\sim$ in $L_e^\sim$ by setting

$$\nabla^\sim \lambda_\alpha^\sim = -ie\hbar^{-1}\mathscr{A}_\alpha \otimes \lambda_\alpha^\sim . \tag{10.23}$$

It follows from Eq. (10.14) that the curvature form of the connection $\nabla^\sim$ in $L_e^\sim$ is given by $-e\hbar^{-1}f$. Since the transition functions $c_{\alpha\beta}$ have modulus one and $\langle\lambda_\alpha^\sim,\lambda_\alpha^\sim\rangle^\sim = 1$, the Hermitian form $\langle,\rangle^\sim$ on $L_e^\sim$ is $\nabla^\sim$-invariant.

The polarization preserving subalgebra of the Poisson algebra of $(\mathscr{T}^*Y,\omega_e)$ consists of functions of the form $q + p_\zeta$, where $q = \check{q}\circ\pi$ is constant along D and p_ζ is defined by Eq. (7.42). The quantization of such functions is given by Eq. (6.24). For $q = \check{q}\circ\pi$ the corresponding operator is given by

$$\mathscr{Q}q[\lambda \otimes \nu_g] = q\lambda \otimes \nu_g . \tag{10.24}$$

For a function p_ζ, Eq. (6.24) and the discussion in Sec. 7.2 leading from Eq. (7.77) to Eq. (7.82) yield

$$\mathscr{Q}p_\zeta[\lambda \otimes \nu_g] = [\{-i\hbar\nabla_{\xi_{p_\zeta}} + p_\zeta - \tfrac{1}{2}i\hbar(\mathrm{Div}\ \zeta)\circ\pi\}\ \lambda] \otimes \nu_g . \tag{10.25}$$

If support $\lambda \subseteq \pi^{-1}(V_\alpha)$, then we can express $\lambda \otimes \nu_g$ in terms of the local section $\lambda_\alpha \otimes \nu_g$ as $\Psi_\alpha \otimes \lambda_\alpha \otimes \nu_g$, and we obtain, with the help of Eq. (10.15),

$$\mathscr{Q}p_\zeta[\Psi_\alpha\otimes\lambda_\alpha\otimes\nu_g] = \{[-i\hbar(\nabla_\zeta + \tfrac{1}{2}\mathrm{Div}\ \zeta) - e\mathscr{A}_\alpha(\zeta)]\Psi_\alpha\}\otimes\lambda_\alpha\otimes\nu_g . \tag{10.26}$$

Since the representation space $\mathscr{H}_e$ is isomorphic to the space of square integrable sections of $L_e^\sim$ via the correspondence $\lambda \otimes \nu_g \mapsto \lambda^\sim$, any linear operator A on $\mathscr{H}_e$ induces a linear operator $A^\sim$ acting on sections of $L_e^\sim$. Taking into account Eq. (10.23), we see that the operator $(\mathscr{Q}p_\zeta)^\sim$ corresponding to $\mathscr{Q}p_\zeta$ is given by

$$(\mathscr{Q}p_\zeta)^\sim[\lambda^\sim] = -i\hbar(\nabla_\zeta^\sim + \tfrac{1}{2}\operatorname{Div}\zeta)\lambda^\sim. \tag{10.27}$$

Similarly, we have from Eq. (10.24)

$$(\mathscr{Q}p)^\sim[\lambda^\sim] = \check{q}\lambda^\sim \tag{10.28}$$

for each function $q = \check{q}\circ\pi$.

The classical relativistic dynamics of a charged particle in an electromagnetic field is described by the Hamiltonian vector field ξ_N of the mass-squared function

$$N(x) = g(x,x), \tag{10.29}$$

cf. Sec. 2.3. We are going to derive an explicit expression for the quantum operator $\mathscr{Q}N$. We shall see that it corresponds to the Laplace-Beltrami operator of the bundle $L_e^\sim$ modified by the curvature term $\hbar^2R/6$, cf. Eq. (7.114). Thus the Klein-Gordon equation can be interpreted as the equation for determining the eigenvectors of the mass-squared operator.

The local one-parameter group ϕ_N^t generated by ξ_N does not preserve the vertical polarization. Assuming that ξ_N is complete and that, for small $t > 0$, the polarizations F and $\mathscr{T}\phi_N^t(F)$ are transverse, we can use the technique developed in Sec. 6.3 to evaluate $\mathscr{Q}N$. Let σ and σ' be two sections in $\mathscr{H}_e$. The Blattner-Kostant-Sternberg kernel $\mathscr{K}_t\colon \mathscr{H}_e \times \mathscr{H}_{e_t} \to \mathbb{C}$ is given by

$$\mathscr{K}_t(\sigma',\phi_N^t\sigma) = \int_{\mathscr{T}^*Y} \langle\sigma',\phi_N^t\sigma\rangle\, |\omega_e^4|, \tag{10.30}$$

where $\langle\sigma',\phi_N{}^t\sigma\rangle$ is defined by Eq. (5.11). If σ and σ' have supports in $\pi^{-1}(V)$ for some coordinate neighborhood V, there exist functions ψ and ψ' on R^4 with supports in the image of the chart $\check{q}: V \to R^4$ such that

$$\sigma = \psi(\underline{q})\lambda \otimes \nu \quad \text{and} \quad \sigma' = \psi'(\underline{q})\lambda \otimes \nu, \tag{10.31}$$

where λ is a local section of L_e satisfying Eq. (10.15) and ν is given by Eq. (10.3).

In the following we shall work in terms of the coordinates $\check{\underline{q}}: V \to R^4$ chosen above. Hence, we have dropped the index α labelling different charts. Substituting the above expressions for σ and σ' into Eq. (5.11), we can rewrite Eq. (10.30) in the form

$$\begin{aligned}\mathscr{K}_t(\psi'(\underline{q})\lambda\otimes\nu, \phi_N{}^t(\psi(\underline{q})\lambda\otimes\nu)) &= (i/\hbar)^2\int_{\pi^{-1}(V)}\Big\{\psi'(\underline{q})\overline{\psi}(\phi_N{}^t\underline{q}) \\ &\times\left[\det \omega_e(\xi_{q^j}, \phi_N{}^t\xi_{q^k})\right]^{\frac{1}{2}}\langle\lambda,\phi_N{}^t\lambda\rangle\Big\}d^4p^e d^4q,\end{aligned} \tag{10.32}$$

where

$$p^e{}_j := p_j + e\mathscr{A}_j\circ\pi \tag{10.33}$$

and $\mathscr{A}_j$ are the components of the form $\mathscr{A}$ with respect to the basis $d\check{q}^1,\ldots,d\check{q}^4$,

$$\mathscr{A} = \sum_j \mathscr{A}_j\, d\check{q}^j. \tag{10.34}$$

Eqs. (10.33) and (10.34) imply that

$$d^4p^e d^4q = d^4p d^4q. \tag{10.35}$$

Taking into account Eqs. (10.18) and (7.70) we can write the scalar product of the sections σ and σ' given by Eq. (10.31) in the form

$$(\psi'(\underline{q})\lambda\otimes\nu|\psi(\underline{q})\lambda\otimes\nu) = \int_V \psi'(\check{\underline{q}})\overline{\psi}(\check{\underline{q}})d^4\check{q}. \tag{10.36}$$

This enables us to rewrite Eq. (10.32) as

$$\mathcal{K}_t(\psi'(\underline{q})\lambda\otimes\nu, \phi_N{}^t(\psi(\underline{q})\lambda\otimes\nu)) = (\psi'(\underline{q})\lambda\otimes\nu|\psi_t(\underline{q})\lambda\otimes\nu), \tag{10.37}$$

where

$$\psi_t(\check{\underline{q}}(y)) \tag{10.38}$$
$$= (i\hbar)^{-2}\int_{\mathcal{T}_y{}^*Y}\psi(\phi_N{}^t\underline{q})\left[\det\, \omega_e(\xi_{q^j}, \phi_N{}^t\xi_{q^k})\right]^{\frac{1}{2}}\langle\phi_N{}^t\lambda,\lambda\rangle d^4p.$$

According to Eqs. (5.1), (6.28) and (6.29) we have

$$\mathcal{Q}N[\psi(\underline{q})\lambda\otimes\nu] = i\hbar\frac{d}{dt}\psi_t(\underline{q})\lambda\otimes\nu\Big|_{t=0}. \tag{10.39}$$

The argument leading to Eq. (6.48) yields

$$\langle\phi_N{}^t\lambda,\lambda\rangle = \exp\left\{i\hbar^{-1}\int_0^t[(\theta_Y + e\pi^*\mathcal{A})(\xi_N)-N]\circ\phi_N{}^{-s}ds\right\} \tag{10.40}$$

and, taking into account Eqs. (2.14), (2.26) and (10.34), we obtain

$$(\theta_Y + e\pi^*\mathcal{A})(\xi_N) = 2N + 2e\sum_{ij}p_i(g^{ij}\mathcal{A}_j)\circ\pi. \tag{10.41}$$

Hence,

$$\psi_t(\check{\underline{q}}(y)) = (i\hbar)^{-2}\int_{\mathcal{T}_y{}^*Y}\psi(\underline{q}\circ\phi_N{}^{-t})\left[\det\,\omega_e(\xi_{q^j}, \phi_N{}^t\xi_{q^k})\right]^{\frac{1}{2}}$$
$$\times\exp\left\{i\hbar^{-1}\int_0^t\left[N + 2e\sum_{kj}p_k(g^{kj}A_j)\circ\pi\right]\circ\phi_N{}^{-s}ds\right\}d^4p. \tag{10.42}$$

We must now compute the derivative with respect to t of the right hand side of Eq. (10.42) at $t = 0$. This can be done by approximating the integrand so that the integration will give results accurate to the first order in t. From an argument analogous to that in Sec. 7.1, we conclude that it suffices to approximate the integrand up to terms of order 1 in t and order 2 in tp_i. The computations are simplified if we assume that the chart $\check{q}: V \to \mathbf{R}^4$ is normal at y. In this case,

$$g^{ij}(y) = \begin{bmatrix} -1 & 0 & 0 & 0 \\ 0 & -1 & 0 & 0 \\ 0 & 0 & -1 & 0 \\ 0 & 0 & 0 & 1 \end{bmatrix} \tag{10.43}$$

and, for each $x \in \mathscr{T}_y{}^*Y$, we have

$$q^i \circ \phi_N{}^{-t}(x) = -2t \sum_{ij} g^{ij}(y) p_j(x) + \text{higher-order terms} \tag{10.44}$$

$$\left[\det \omega_e(\xi_{q^i}(x), \phi_N{}^t \xi_{q^j}(x))\right]^{\frac{1}{2}} = i(2t)^2$$
$$+ \text{ higher-order terms} \tag{10.45}$$

$$\begin{aligned}[N + 2e \sum_{jk} p_j (g^{jk}\mathscr{A}_k) \circ \pi] \circ \phi_N{}^{-s}(x) &\\ = N(x) + 2e \sum_{jk} p_j(x) g^{jk}(y) \mathscr{A}_k(y) &\\ - 4es \sum_{jkmn} p_j(x) p_k(x) g^{jm}(y) g^{kn}(y) \mathscr{A}_{m,n}(y) &\\ + \text{higher-order terms}, &\end{aligned} \tag{10.46}$$

whoro

$$\mathscr{A}_{m,n} := \frac{\partial}{\partial \check{q}^n} \mathscr{A}_m. \tag{10.47}$$

Integrating Eq. (10.46) with respect to s we obtain

$$\begin{aligned}\int_0^t [N + 2e \sum_{jk} p_j (g^{jk}\mathscr{A}_k) \circ \pi] \circ \phi_N{}^{-s}(x)\,ds &\\ = tN(x) + 2et \sum_{jk} p_j(x) g^{jk}(y) \mathscr{A}_k(y) &\\ - 2et^2 \sum_{jkmn} p_j(x) p_k(x) g^{jm}(y) g^{kn}(y) \mathscr{A}_{m,n}(y) &\\ + \text{higher-order terms}. &\end{aligned} \tag{10.48}$$

Substituting Eqs. (10.43), (10.44), (10.45), (10.46) and (10.48) into Eq. (10.42), we have

$$\begin{aligned}\psi_t(\underline{0}) &= i(2t)^2(i\hbar)^{-2}\int_{\mathscr{T}_y^*Y}\Big\{\psi(2tp_1,2tp_2,2tp_3,-2tp_4)\\ &\times \exp[it\hbar^{-1}(-p_1{}^2-p_2{}^2-p_3{}^2+p_4{}^2)]\\ &\times \exp[2iet\hbar^{-1}\sum_{jk} p_j g^{jk}(y)\mathscr{A}_k(y)]\\ &\times \exp[-2iet^2\hbar^{-1}\sum_{jkmn} p_jp_k g^{jm}(y)g^{kn}(y)\mathscr{A}_{m,n}(y)]\Big\}d^4p\\ &+ \text{higher-order terms.}\end{aligned} \tag{10.49}$$

Changing the variables of integration from p_i to

$$u_i = 2tp_i|\mathscr{T}_y^*Y \tag{10.50}$$

and taking into account Eqs. (7.26) and (7.27), we obtain

$$\begin{aligned}\lim_{t\to 0+}\psi_t(\underline{0}) &= i\hbar\Big(-\frac{\partial^2}{\partial u_2{}^2}-\frac{\partial^2}{\partial u_2{}^2}-\frac{\partial^2}{\partial u_3{}^2}+\frac{\partial^2}{\partial u_4{}^2}\Big)\\ &\Big\{\psi(u_1,u_2,u_3,-u_4)\exp\Big[ie\hbar^{-1}\sum_{jk}u_jg^{jk}(y)\mathscr{A}_k(y)\Big]\\ &\times\exp\Big[-\tfrac{1}{2}ie\hbar^{-1}\sum_{jkmn}u_ju_kg^{jm}(y)g^{kn}(y)\mathscr{A}_{m,n}(y)\Big]\Big\}\Big|_{\underline{u}=0},\end{aligned} \tag{10.51}$$

which can be written

$$\begin{aligned}\lim_{t\to 0+}&\psi_t(\underline{0})\\ &= i\hbar\sum_{mn}g^{mn}(y)\Big[\Big(\frac{\partial^2}{\partial u_m\partial u_n}-2ie\hbar^{-1}\mathscr{A}_m(y)\frac{\partial}{\partial u_m}\Big)\psi(u_1,u_2,u_3,u_4)\Big]_{\underline{u}=0}\\ &+ i\hbar\sum_{mn}g^{mn}(y)\Big[-e^2\hbar^{-2}\mathscr{A}_m(y)\mathscr{A}_n(y) - ie\hbar^{-1}\mathscr{A}_{m,n}(y)\Big]\psi(\underline{0}).\end{aligned} \tag{10.52}$$

Introducing the function Ψ on V related to ψ by Eq. (7.88), we have

$$\pm\Psi\otimes\lambda\otimes\nu_g = \psi(\underline{q})\lambda\otimes\nu \tag{10.53}$$

and can write the term on the right hand side of Eq. (10.52), which contains second derivations of ψ, in the covariant form

$$\sum_{mn} g^{mn}(y)\left[\frac{\partial^2}{\partial u_m \partial u_n} \psi(u_1,u_2,u_3,u_4)\right]\Big|_{u=0}$$
$$= \pm|\det g(y)|^{\frac{1}{4}}\left\{\left[\sum_{mn} g^{mn}(y)\nabla_m\nabla_n - \frac{1}{6} R(y)\right]\Psi\right\}(y). \tag{10.54}$$

Similarly, the covariant derivative $\nabla_n \mathscr{A}$ of the form $\mathscr{A}$ with respect to the vector field $\frac{\partial}{\partial \check{q}^n}$ is given at y by

$$(\nabla_n \mathscr{A})(y) = \sum_m \mathscr{A}_{m,n}(y) d\check{q}^m \tag{10.55}$$

since $\Gamma_{mn}{}^k(y) = 0$. Making use of Eqs. (10.54) and (10.55), we can write

$$\lim_{t\to 0+} \frac{d}{dt} \psi_t(\underline{0}) = \pm i\hbar|\det g(y)|^{\frac{1}{4}}$$
$$\times \left\{\left[\sum_{mn} g^{mn}(\nabla_m - ie\hbar^{-1}\mathscr{A}_m)(\nabla_n - ie\hbar^{-1}\mathscr{A}_n) - \frac{1}{6}R\right]\Psi\right\}(y). \tag{10.56}$$

Since

$$\mathscr{Q}N[\Psi\otimes\lambda\otimes\nu_g] - i\hbar \frac{d}{dt}\psi_t(\underline{q})\lambda\otimes\nu\Big|_{t=0} \tag{10.57}$$

for each $x \in \mathscr{T}_y^*Y$, Eqs. (7.88), (10.53) and (10.56) yield

$$[(\mathscr{Q}N)(\Psi\otimes\lambda\otimes\nu_g)](x)$$
$$= \left\{-\hbar^2\left[\sum_{mn} g^{mn}(\nabla_m - ie\hbar^{-1}\mathscr{A}_m)(\nabla_n - ie\hbar^{-1}\mathscr{A}_n) - \frac{1}{6}R\right]\Psi\otimes\lambda\otimes\nu_g\right\}(x). \tag{10.58}$$

The expression in the square bracket on the right hand side of Eq. (10.58) is an invariant differential operator which acts on the function Ψ with support in V. Hence, Eq. (10.58) is valid for all domains V and all $x \in \pi^{-1}(V)$, and can be rewritten in a global form in terms of the connection $\nabla^{\sim}$ in the bundle $L_e{}^{\sim}$. The operator $\mathscr{Q}N$ defines a linear operator $(\mathscr{Q}N)^{\sim}$ acting on the sections $\lambda^{\sim}$ of $L_e{}^{\sim}$ as follows:

$$(\mathscr{Q}N)^{\sim}[\lambda^{\sim}] = -\hbar^2(\Delta^{\sim} - \frac{1}{6} R)\lambda^{\sim} \tag{10.59}$$

where $\Delta^{\sim}$ denotes the Laplace-Beltrami operator defined in terms of the metric g on Y and the connection $\nabla^{\sim}$ in $L_e{}^{\sim}$. Eq. (10.59) is a direct consequence of Eqs. (10.23) and (10.58).

10.2. Charge superselection rules

The formulation of relativistic dynamics presented in the previous section treats the mass and the charge asymmetrically. The mass is a dynamical variable while the charge is a fixed parameter in the theory. This asymmetry disappears in the five-dimensional theory of Kaluza. One obtains charge quantization in the generalization of the Kaluza theory which is briefly outlined below.

In the generalized Kaluza theory the phase space of a particle in external gravitational and electromagnetic fields is $(\mathscr{T}^*Z, d\theta_Z)$, where Z is a T^1-principal fibre bundle over the space-time manifold Y. The bundle Z is endowed with a T^1-invariant metric k of signature $(-,-,-,+,+)$ such that the fundamental vector field η_1, corresponding to the real number 1 in the Lie algebra of T^1, has constant length 1, that is,

$$\mathscr{L}_{\eta_1} k = 0 \tag{10.60}$$

and

$$k(\eta_1,\eta_1) = 1. \tag{10.61}$$

We identify T^1 with the multiplicative group of complex numbers of modulus 1. The Lie algebra of T^1 is identified with $\mathbb{R}$ by associating, to each $r \in \mathbb{R}$, the one-parameter group $t \mapsto \exp(2\pi i e_0 h^{-1} rt)$, where e_0 is a parameter interpreted as the elementary charge.

Let α be the 1-form on Z defined by

$$\alpha(u) = k(\eta_1(z), u) \tag{10.62}$$

for each $z \in Z$ and each $u \in \mathscr{T}_z Z$. Eqs. (10.60) and (10.61) imply that, for each fundamental vector field η_r on Z,

$$\mathscr{L}_{\eta_r} \alpha = 0 \quad \text{and} \quad \alpha(\eta_r) = r. \tag{10.63}$$

Hence α is the connection form of a connection in Z. The curvature form of this connection defines a closed 2-form f on Y such that

$$d\alpha = \varkappa^* f, \tag{10.64}$$

where $\varkappa: Z \to Y$ denotes the T^1-principal fibre bundle projection. The horizontal distribution $\operatorname{hor}\mathscr{T}Z$ on Z defined by

$$\operatorname{hor}\mathscr{T}Z = \{u \in \mathscr{T}Z \,|\, \alpha(u) = 0\} \tag{10.65}$$

is orthogonal to the vertical distribution tangent to the fibres of $\varkappa$. Hence, k decomposes into vertical and horizontal parts

$$k = \operatorname{ver} k + \operatorname{hor} k \tag{10.66}$$

where, for each $u, v \in \mathscr{T}_z Z$,

$$\operatorname{ver} k(u,v) = k(\operatorname{ver} u, \operatorname{ver} v) \tag{10.67}$$

and

$$\operatorname{hor} k(u,v) = k(\operatorname{hor} u, \operatorname{hor} v). \tag{10.68}$$

Since k is T^1-invariant its horizontal part is also T^1-invariant and there exists a unique metric g on Y such that

$$\text{hor } k = \varkappa^* g. \tag{10.69}$$

Eq. (10.61) implies that g has the signature $(-,-,-,+)$.

Each $x \in \mathcal{T}_z^*Z$ can be decomposed into its vertical and horizontal parts

$$x = \text{ver } x + \text{hor } x \tag{10.70}$$

where, for each $u \in \mathcal{T}_z Z$,

$$(\text{ver } x)(u) = x(\text{ver } u) \tag{10.71}$$

and

$$(\text{hor } x)(u) = x(\text{hor } u). \tag{10.72}$$

Substituting $\eta_1(z)$ for u in Eq. (10.71) and taking into account Eq. (7.42) we obtain

$$(\text{ver } x)(\eta_1(z)) = p_{\eta_1}(x). \tag{10.73}$$

Since η_1 spans the vertical distribution on Z and, according to Eq. (10.63), $\alpha(\eta_1) = 1$, we have

$$\text{ver } x = p_{\eta_1}(x)\alpha_z. \tag{10.74}$$

Let $\hat{\varkappa}: \mathcal{T}^*Z \to \mathcal{T}^*Y$ be the mapping defined by

$$\hat{\varkappa}(x)(\mathcal{T}\varkappa(u)) = (\text{hor } x)(u) \tag{10.75}$$

for each $x \in \mathcal{T}^*Z$ and each $u \in \mathcal{T}_{\pi(x)}Z$. It is a submersion such that the diagram

$$\begin{array}{ccc} \mathcal{T}^*Z & \xrightarrow{\hat{\varkappa}} & \mathcal{T}^*Y \\ \pi \downarrow & & \downarrow \pi \\ Z & \xrightarrow{\varkappa} & Y \end{array}, \tag{10.76}$$

in which the vertical arrows denote the cotangent bundle projections, commutes.

The physical interpretation of the theory follows from the identification of Y with the space-time manifold, g with the gravitational field, and f with the electromagnetic field. The canonical momentum p_{η_1} in the η_1 direction is identified with the charge Q. That is, for each $x \in \mathscr{T}^*Z$, the *charge in the classical state* x is

$$Q(x) = p_{\eta_1}(x). \tag{10.77}$$

For each function $f: \mathscr{T}^*Y \to \mathbf{R}$, the pull-back

$$\hat{f} = f \circ \varkappa \tag{10.78}$$

of f to $\mathscr{T}^*Z$ is interpreted as the dynamical variable in the Kaluza theory corresponding to f. In particular, for each $x \in \mathscr{T}^*Z$,

$$\hat{N}(x) = N \circ \hat{\varkappa}(x) = k(\text{hor } x, \text{hor } x) \tag{10.79}$$

is the square of the mass in the classical state x. As before, we use the same symbol to denote both the metric and the induced bilinear form on the cotangent bundle.

The Lagrange bracket is the exterior differential of the canonical 1-form θ_Z on $\mathscr{T}^*Z$. Taking into account Eqs. (2.12), (10.70), (10.74), (10.75) and (10.76), we can decompose θ_Z as follows:

$$\theta_Z = \hat{\varkappa}^*\theta_Y + Q\pi^*\alpha, \tag{10.80}$$

where $\pi^*\alpha$ denotes the pull-back to $\mathscr{T}^*Z$ of the connection form α on Z. Differentiating Eq. (10.80), and substituting $\varkappa^*f$ for $d\alpha$ according to Eq. (10.64), we obtain

$$d\theta_Z = \hat{\varkappa}^*d\theta_Y + Q(\varkappa \circ \pi)^*f + dQ \wedge \pi^*\alpha, \tag{10.81}$$

where $(\varkappa \circ \pi)^* f$ is the pull-back to $\mathcal{T}^* Z$ of the electromagnetic field f on Y. Since the diagram (10.76) commutes, we can rewrite Eq. (10.81) in the form

$$d\theta_Z = \hat{\varkappa}^* d\theta_Y + Q\hat{\varkappa}^* \pi^* f + dQ \wedge \pi^* \alpha. \tag{10.82}$$

Hence, for each function $\hat{f}$ of the form given by Eq. (10.78) we have

$$[\hat{f}, Q] = 0. \tag{10.83}$$

Consider the Hamiltonian vector field $\xi_{\hat{N}}$ of the mass-squared function $\hat{N}$ defined by Eq. (10.79). Eq. (10.83) implies that

$$\xi_{\hat{N}} Q = 0 \tag{10.84}$$

so that, for each value of the charge $e \in R$, the vector field $\xi_{\hat{N}}$ restricts to a vector field $\xi_{\hat{N}} | Q^{-1}(e)$ on $Q^{-1}(e)$. Eq. (10.82) implies that $\hat{\varkappa}$ restricted to $Q^{-1}(e)$ maps $Q^{-1}(e)$ onto $\mathcal{T}^* Y$ in such a way that $d\theta_Z | Q^{-1}(e)$ is the pull-back of the form ω_e given by Eq. (10.1). Moreover, $\mathcal{T}\hat{\varkappa}$ maps $\xi_{\hat{N}} | Q^{-1}(e)$ onto the Hamiltonian vector field ξ_N on $\mathcal{T}^* Y$ defined with respect to the symplectic form ω_e which describes the dynamics of a particle with charge e. Hence, $\xi_{\hat{N}}$ describes the dynamics of particles with variable charge.

Let $\hat{D}$ be the distribution on $\mathcal{T}^* Z$ spanned by the Hamiltonian vector field of Q and the Hamiltonian vector fields of the pull-backs to $\mathcal{T}^* Z$ of functions on Y; it is an involutive Lagrangian distribution on $(\mathcal{T}^* Z, d\theta_Z)$. The integral curves of ξ_Q are periodic, and thus the integral manifolds of $\hat{D}$ are diffeomorphic to $R^4 \times T^1$. The space $\mathcal{T}^* Z / \hat{D}$ of integral manifolds of $\hat{D}$ is a quotient manifold of $\mathcal{T}^* Z$ with projection denoted by $\pi_{\hat{D}}: \mathcal{T}^* Z \to \mathcal{T}^* Z / \hat{D}$. Moreover, $\mathcal{T}^* Z / \hat{D}$ is is diffeomorphic to the product $Y \times \mathbf{R}$ so that the following diagram commutes

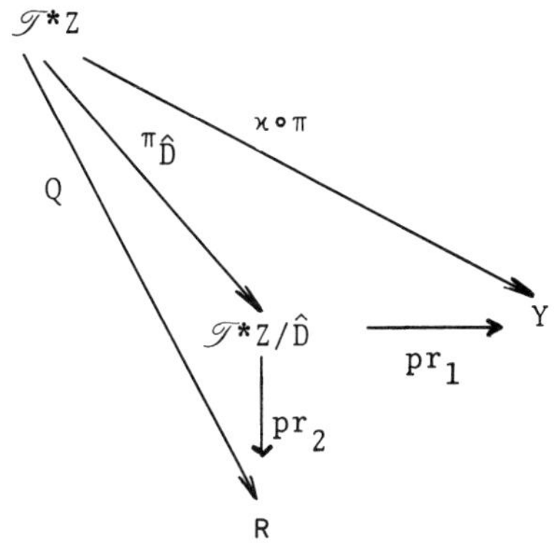

. (10.85)

The complexification $\hat{F} = \hat{D}^{\mathbf{C}}$ of $\hat{D}$ is a complete strongly admissible real polarization of $(\mathscr{T}^*Z, d\theta_Z)$. Using the notation of Sec. 4.5 we denote by K the distribution spanned by the vector field ξ_Q. The integral manifolds of K are diffeomorphic to T^1 and there is a unique density κ on K which assigns a total length 1 to each integral circle of K. The relation between the polarization $\hat{F}$ on $\mathscr{T}^*Z$ and the vertical polarization F on $\mathscr{T}^*Y$ is given by

$$\mathscr{T}\hat{\chi}(\hat{F}) = F \tag{10.86}$$

and

$$(\ker \mathscr{T}\hat{\chi}) \cap \hat{F} = K^{\mathbf{C}}. \tag{10.87}$$

If Y is orientable then so is Z and $(\mathscr{T}^*Z, d\theta_Z)$ admits a metaplectic structure which induces a metalinear frame bundle $\tilde{\mathscr{B}}\hat{F}$ of $\hat{F}$. We denote by $\sqrt{\wedge^5 \hat{F}}$ the line bundle associated to $\tilde{\mathscr{B}}\hat{F}$ corresponding to the character χ of $ML(5,\mathbf{C})$. For each chart $\check{\underline{q}}_\alpha : V_\alpha \to \mathbf{R}^4$ on Y, the linear frame field

$$\underline{\hat{\xi}}^{\alpha} = (\xi_{\hat{\underline{q}}_\alpha}, \xi_Q), \qquad (10.88)$$

where $\hat{\underline{q}}_\alpha = \check{\underline{q}}_\alpha \circ \varkappa \circ \pi$, trivializes $\mathscr{B}\hat{F}|(\varkappa\circ\pi)^{-1}(V_\alpha)$. The restriction of $\tilde{\mathscr{B}}\hat{F}$ to $(\varkappa\circ\pi)^{-1}(V_\alpha)$ is also trivial, and we denote by $\tilde{\underline{\hat{\xi}}}^{\alpha}$ a section of $\tilde{\mathscr{B}}\hat{F}$ projecting onto $\underline{\hat{\xi}}^{\alpha}$. Let $\hat{\nu}_\alpha$ be the section of $\sqrt{\wedge}^5\hat{F}$ defined by

$$\hat{\nu}_\alpha{}^{\#} \circ \tilde{\underline{\hat{\xi}}}^{\alpha} = 1. \qquad (10.89)$$

The orientability of Y implies, in the same way as in Sec. 7.2, the existence of a global section $\hat{\nu}_g$ of $\sqrt{\wedge}^5\hat{F}$ such that

$$\hat{\nu}_g|(\varkappa\circ\pi)^{-1}(V_\alpha) = \pm|(\det g_\alpha)\circ\varkappa\circ\pi|^{\frac{1}{4}}\nu_\alpha, \qquad (10.90)$$

where g_α is given by Eq. (7.57). Clearly, $\hat{\nu}_g$ is covariantly constant along $\hat{F}$.

Let $\hat{L}$ denote the trivial complex line bundle over $\mathscr{T}^*Z$ with a trivializing section $\hat{\lambda}_o$, a connection ∇ such that

$$\nabla\hat{\lambda}_o = -i\hbar^{-1}\theta_Z \otimes \hat{\lambda}_o \qquad (10.91)$$

and a Hermitian form $\langle,\rangle$ satisfying

$$\langle\hat{\lambda}_o, \hat{\lambda}_o\rangle = 1. \qquad (10.92)$$

The section $\hat{\lambda}_o$ is not covariantly constant along $\hat{F}$; Eqs. (10.77), (10.91) and (7.44) yield

$$\nabla_{\xi_Q}\hat{\lambda}_o = -i\hbar^{-1}Q\hat{\lambda}_o. \qquad (10.93)$$

Since the one-parameter group in T^1 generated by $1 \in R$ is given by $t \mapsto \exp(2\pi i e_o h^{-1} t)$, the integral curves of the fundamental vector field η_1 on Z are periodic with period $he_o{}^{-1}$. Therefore, the integral curves of ξ_Q are periodic with period $\hbar e_o^{-1}$. For each $x \in \mathscr{T}^*Z$, the holonomy group at x of $L \otimes \sqrt{\wedge^5 \hat{F}} | \Lambda$, where Λ is the integral manifold of $\hat{D}$ through x, is therefore generated by $\exp(2\pi i e_o^{-1} Q(x))$. Hence, the Bohr-Sommerfeld set S in $\mathscr{T}^*Z$ is given by

$$S = \{x \in \mathscr{T}^*Z \mid e_o^{-1} Q(x) \in Z\}. \tag{10.94}$$

For each $n \in Z$, let S_n be the subset of S given by

$$S_n = Q^{-1}(ne_o) \tag{10.95}$$

and $\mathscr{H}_n$ the subspace of the representation space $\mathscr{H}$ consisting of sections with supports in S_n. Since Q is constant along $\hat{F}$, Eqs. (6.26) and (10.95) imply that $\mathscr{H}_n$ is the eigenspace of the charge operator $\mathscr{Q}Q$ with eigenvalue ne_o. The representation space is the direct sum of the $\mathscr{H}_n$:

$$\mathscr{H} = \bigoplus_{n \in Z} \mathscr{H}_n. \tag{10.96}$$

According to the argument given in Sec. 6.4, for each dynamical variable f on $\mathscr{T}^*Z$ quantized in the way described in Chapter 6, the operators $\mathscr{Q}f$ and $\mathscr{Q}Q$ commute,

$$[\mathscr{Q}f, \mathscr{Q}Q] = 0. \tag{10.97}$$

Hence, $\mathscr{Q}f$ is completely specified by the collection $\{\mathscr{Q}_n f \mid n \in Z\}$ of operators on $\mathscr{H}_n$, where $\mathscr{Q}_n f$ is the restriction of $\mathscr{Q}f$ to $\mathscr{H}_n$,

$$\mathscr{Q}_n f = \mathscr{Q}f \,|\, \mathscr{H}_n. \tag{10.98}$$

10.3. Quantization in the Kaluza theory

Given an integer n, we are going to construct a unitary map $\mathcal{U}_n: \mathcal{H}_n \to \mathcal{H}_e$, where

$$e = ne_o, \tag{10.99}$$

such that

$$\mathcal{U}_n \mathcal{Q}_n(f \circ \hat{\varkappa}) = \mathcal{Q}f \, \mathcal{U}_n \tag{10.100}$$

for all the dynamical variables $f: \mathcal{T}^*Y \to \mathbb{R}$ quantized in Sec. 10.1. This will guarantee the equivalence of the two approaches to the quantum dynamics of a charged particle presented in this chapter.

The space $\mathcal{H}_n$ consists of sections of $\hat{L} \otimes \sqrt{\wedge^5 \hat{F}}|S_n$ which are covariantly constant along $\hat{F}|S_n$ and square integrable over $\pi_{\hat{D}}(S_n)$, where S_n is the component of the Bohr-Sommerfeld variety given by Eq. (10.95). Since $\sqrt{\wedge^5 \hat{F}}$ has a nonvanishing covariantly constant section $\hat{\nu}_g$, every section $\sigma \in \mathcal{H}_n$ can be written in the form

$$\sigma = \hat{\lambda} \otimes \hat{\nu}_g, \tag{10.101}$$

where $\hat{\lambda}$ is a section of $\hat{L}|S_n$ covariantly constant along $\hat{F}|S_n$.

For each $x \in S_n$, the restriction of $\hat{L}$ to the fibre $\hat{\varkappa}^{-1}(\hat{\varkappa}(x))$ of $\hat{\varkappa}: \mathcal{T}^*Z \to \mathcal{T}^*Y$ through x has a flat connection with a trivial holonomy group. Hence, there is an equivalence relation $\sim$ in $\hat{L}|S_n$ given by $\hat{\ell}_1 \sim \hat{\ell}_2$ if and only if $\hat{\ell}_1$ can be joined to $\hat{\ell}_2$ by a horizontal curve contained in $\hat{L}|\hat{\varkappa}^{-1}(\hat{\varkappa}(x))$ for some $x \in S_n$. The space $\hat{L}_n$ of $\sim$ equivalence classes in $\hat{L}|S_n$ has the structure of a complex line bundle over $\mathcal{T}^*Y$. The canonical projection $\hat{\varkappa}_n: \hat{L}|S_n \to \hat{L}_n$ is an isomorphism on each fibre and the following diagram

$$\begin{array}{ccc} \hat{L}|S_n & \xrightarrow{\hat{\varkappa}_n} & \hat{L}_n \\ \downarrow & & \downarrow \\ S_n & \xrightarrow[\hat{\varkappa}|S_n]{} & \mathscr{T}^*Y \end{array}, \tag{10.102}$$

in which the vertical arrows denote the line bundle projections, commutes. The connection in $\hat{L}$ induces a connection in $\hat{L}_n$, the curvature form of which can be obtained by restricting the curvature form of $\hat{L}$ to S_n. Eqs. (10.82) and (10.99) yield

$$d\theta_Z|S_n = (\hat{\varkappa}|S_n)^*(d\theta_Y + e\pi^* f). \tag{10.103}$$

Thus, the curvature form of the connection in $\hat{L}_n$ is the same as the curvature form of the connection in the bundle L_e introduced in Sec. 10.1. Hence, we can identify L_e with $\hat{L}_n$:

$$\hat{L}_n = L_e. \tag{10.104}$$

Each section $\hat{\lambda}$ of $\hat{L}|S_n$ covariantly constant along $\hat{P}|S_n$ gives rise to a unique section λ of L_e such that, for each $x \in S_n$,

$$\hat{\varkappa}_n(\hat{\lambda}(x)) = \lambda(\hat{\varkappa}(x)). \tag{10.105}$$

Similarly, there is a bundle map $\hat{\varkappa}_n\colon \sqrt{\wedge}^5\hat{F} \to \sqrt{\wedge}^4 F$ such that

$$\hat{\varkappa}_n(\hat{\nu}_g(x)) = \nu_g(\hat{\varkappa}(x)) \tag{10.106}$$

for each $x \in S_n$. Since $\hat{\nu}_g$ and ν_g are nonvanishing sections of $\sqrt{\wedge}^5\hat{F}$ and $\sqrt{\wedge}^4 F$, respectively, $\hat{\varkappa}_n\colon \sqrt{\wedge}^5\hat{F} \to \sqrt{\wedge}^4 F$ is an isomorphism on each fibre.

Let $\mathscr{U}_n: \mathscr{A}_n \to \mathscr{A}_e$ be defined as follows. For each $x \in S_n$, set

$$\mathscr{U}_n(\hat{\lambda} \otimes \hat{\nu}_g)(\hat{\varkappa}(x)) = (e_o h^{-1})^{\frac{1}{2}} \hat{\varkappa}_n(\hat{\lambda}(x)) \otimes \hat{\varkappa}_n(\hat{\nu}_g(x)). \quad (10.107)$$

Since $\hat{\lambda}(x)$ and $\hat{\nu}_g(x)$ are covariantly constant along $\hat{F}$, the left hand side of Eq. (10.107) does not depend on the choice of $x \in \hat{\varkappa}^{-1}(\hat{\varkappa}(x))$, and we obtain a section $\mathscr{U}_n(\hat{\lambda} \otimes \hat{\nu}_g)$ of $L_e \otimes \sqrt{\wedge^4 F}$ which is covariantly constant along F. Since $\hat{\varkappa}_n$ induces isomorphisms of the corresponding fibres, the linear map $\mathscr{U}_n: \mathscr{A}_n \to \mathscr{A}_e$ defined by Eq. (10.107) is a vector space isomorphism.

The scalar product on $\mathscr{A}_n$ is given by Eq. (4.46) with $k = 1$ and a single manifold $Q_i = \pi_{\hat{D}}(S_n)$. Since the restriction of $pr_1: \mathscr{T}^*Z/\hat{D} \to Y$ to $\pi_{\hat{D}}(S_n)$ is a diffeomorphism [cf. diagram (10.85) and Eq. (10.95)], we may identify Q_i with Y,

$$Q_i = \pi_{\hat{D}}(S_n) = Y. \quad (10.108)$$

Let $\check{q}_\alpha: V_\alpha \to R^4$ be a local chart on Y and $(\underline{p}_\alpha, \underline{q}_\alpha): \pi^{-1}(V_\alpha) \to R^8$ the induced chart on $\mathscr{T}^*Y$. Consider a local frame field on $\mathscr{T}^*Z$ of the form

$$\left(h e_o^{-1} \xi_Q, \frac{\partial}{\partial \hat{p}_{\alpha_1}}, \ldots, \frac{\partial}{\partial \hat{p}_{\alpha_4}}; \eta, \frac{\partial}{\partial \hat{q}_\alpha^{\;1}}, \ldots, \frac{\partial}{\partial \hat{q}_\alpha^{\;4}}\right), \quad (10.109)$$

where η is a local vector field on $\mathscr{T}^*Z$ chosen so that the frame field (10.109) is symplectic. This frame field satisfies the conditions (4.47), (4.48) and (4.49). The last four vector fields in (10.109) project onto a local linear frame field

$$\left(\frac{\partial}{\partial \check{q}_\alpha^{\;1}}, \ldots, \frac{\partial}{\partial \check{q}_\alpha^{\;4}}\right) \quad (10.110)$$

for Y. Given two sections $\hat{\lambda}_1 \otimes \hat{\nu}_g$ and $\hat{\lambda}_2 \otimes \hat{\nu}_g$ in $\mathscr{A}_n$, the value of the density $<\hat{\lambda}_1 \otimes \hat{\nu}_g, \hat{\lambda}_2 \otimes \hat{\nu}_g>_Y$ on the linear frame field (10.110) is

$$<\hat{\lambda}_1 \otimes \nu_g, \hat{\lambda}_2 \otimes \hat{\nu}_g>_Y\left(\frac{\partial}{\partial \check{q}_\alpha{}^1}, \ldots, \frac{\partial}{\partial \check{q}_\alpha{}^4}\right) = e_o h^{-1} <\lambda_1{}^\sim, \lambda_2{}^\sim> |\det g_\alpha|^{\frac{1}{2}}, \tag{10.111}$$

according to Eqs. (4.51), (10.11), (10.12), (10.89), (10.90) and (10.105). Hence, the scalar product on $\mathscr{A}_n$ is given by

$$(\hat{\lambda}_1 \otimes \hat{\nu}_g | \hat{\lambda}_2 \otimes \hat{\nu}_g) = e_o h^{-1} \int_Y <\lambda_1{}^\sim, \lambda_2{}^\sim> |\det g|^{\frac{1}{2}}. \tag{10.112}$$

This and Eq. (10.13) imply that $\mathscr{U}_n : \mathscr{A}_n \to \mathscr{A}_e$ defined by Eq. (10.107) preserves the scalar product; hence $\mathscr{U}_n$ is unitary.

Let f be a function on $(\mathscr{T}^*Y, \omega_e)$ such that $\phi_f{}^t$ preserves F. Then, the one-parameter group $\phi_{\hat{f}}{}^t$ of canonical transformations of $(\mathscr{T}^*Z, d\theta_Z)$ generated by $\hat{f} = f \circ \hat{\varkappa}$ preserves $\hat{F}$ so that

$$\phi_{\hat{f}}{}^t(S_n) = S_n \tag{10.113}$$

for all $t \in \mathbb{R}$. The following diagram commutes

$$\begin{array}{ccc} S_n & \xrightarrow{\phi_{\hat{f}}{}^t} & S_n \\ \hat{\varkappa} \downarrow & & \downarrow \hat{\varkappa} \\ \mathscr{T}^*Y & \xrightarrow{\phi_f{}^t} & \mathscr{T}^*Y \end{array} \tag{10.114}$$

and, for each $x \in S_n$,

$$\hat{\varkappa}_n[(\phi_{\hat{f}}{}^t \hat{\nu}_g)(x)] = (\phi_f{}^t \nu_g)(\hat{\varkappa}(x)) \tag{10.115}$$

and

$$\hat{\varkappa}_n[(\phi_{\hat{f}}{}^t \hat{\lambda})(x)] = (\phi_f{}^t \lambda)(\hat{\varkappa}(x)), \tag{10.116}$$

where the sections $\hat{\lambda}, \hat{\nu}_g$ and λ, ν_g are related by Eqs. (10.105) and (10.106). Hence,

$$\hat{x}_n\{[\phi_{\hat{f}}{}^t(\hat{\lambda} \otimes \hat{\nu}_g)](x)\} = [\phi_f{}^t(\lambda \otimes \nu_g)](\hat{x}(x)). \tag{10.117}$$

Differentiating Eq. (10.117) with respect to t and setting $t = 0$, we obtain

$$\hat{x}_n\{[\mathscr{Q}_n\hat{f}(\hat{\lambda} \otimes \hat{\nu}_g)](x)\} = [\mathscr{Q}f(\lambda \otimes \nu_g)](\hat{x}(x)). \tag{10.118}$$

Taking into account Eq. (10.107), we can rewrite Eq. (10.118) in the form

$$\mathscr{U}_n\mathscr{Q}_n\hat{f}[\hat{\lambda} \otimes \hat{\nu}_g] = \mathscr{Q}f\,\mathscr{U}_n[\hat{\lambda} \otimes \hat{\nu}_g]. \tag{10.119}$$

Since Eq. (10.119) holds for all sections $\hat{\lambda} \otimes \hat{\nu}_g$ in $\mathscr{H}_n$, it follows that

$$\mathscr{U}_n(\mathscr{Q}_n\hat{f}) = (\mathscr{Q}f)\mathscr{U}_n. \tag{10.120}$$

Thus, the unitary operator $\mathscr{U}_n: \mathscr{H}_n \to \mathscr{H}_e$ intertwines the quantizations of the polarization preserving dynamical variables on $(\mathscr{T}^*Y, \omega_e)$. In particular, for a position function $q = \check{q}\circ\pi$, where $\check{q}$ is a function on Y, we obtain

$$\mathscr{Q}_n(q\circ\hat{x}) = \mathscr{U}_n{}^{-1}(\mathscr{Q}q)\,\mathscr{U}_n, \tag{10.121}$$

and for the momentum p_ζ associated to a vector field ζ on Y we have

$$\mathscr{Q}_n(p_\zeta\circ\hat{x}) = \mathscr{U}_n{}^{-1}(\mathscr{Q}p_\zeta)\,\mathscr{U}_n. \tag{10.122}$$

It remains to verify that Eq. (10.120) holds when f is replaced by the mass-squared function $\hat{N}$. The one-parameter group $\phi_{\hat{N}}{}^t$ of canonical transformations of $(\mathscr{T}^*Z, d\theta_Z)$ generated by $\hat{N}$ does not preserve $\hat{F}$. Since $\hat{N}$ commutes with Q and ξ_Q is contained in $\hat{F}$, the polarizations $\hat{F}$ and $\mathscr{T}\phi_{\hat{N}}{}^t(\bar{\hat{F}})$ intersect along the one-dimensional distribution

$K^{\mathbf{C}}$ spanned by ξ_Q. The distribution K plays the role of the distribution D_{12} of Sec. 5.2, where F_1 and F_2 are identified with $\hat{F}$ and $\mathscr{T}\phi_{\hat{N}}{}^{t}(\hat{F})$ respectively. The space $\mathscr{T}^*Z/K$ of integral manifolds of K is diffeomorphic to $\mathbf{R} \times \mathscr{T}^*Y$ and the diagram

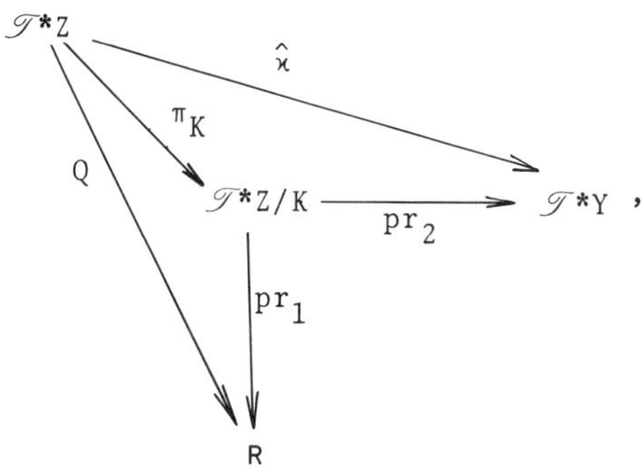

(10.123)

where $\pi_K\colon \mathscr{T}^*Z \to \mathscr{T}^*Z/K$ is the canonical projection, commutes. Hence, $\mathscr{T}^*Z/K$ is a quotient manifold of $\mathscr{T}^*Z$ which implies that the polarizations $\hat{F}$ and $\mathscr{T}\phi_{\hat{N}}{}^{t}(\hat{F})$ form a strongly admissible pair. We can therefore use the technique developed in Sec. 5.2. to evaluate the Blattner-Kostant-Sternberg kernel $\mathscr{K}_t\colon \mathscr{A} \times \mathscr{A}_t \to \mathbf{C}$, where we denote by $\mathscr{A}_t$ the representation space defined by the polarization $\mathscr{T}\phi_{\hat{N}}{}^{t}(\hat{F})$.

Let $\hat{\sigma}_1$ and $\hat{\sigma}_2$ be two eigenvectors of the charge operator $\mathscr{Q}Q$ with eigenvalues $n_1 e_o$ and $n_2 e_o$ respectively. Since $\hat{N}$ and Q commute, $\phi_{\hat{N}}{}^{t}$ preserves each component S_n of S, and $\mathscr{K}_t(\hat{\sigma}_1, \phi_{\hat{N}}{}^{t}\hat{\sigma}_2) = 0$ if $n_1 \neq n_2$. If $n_1 = n_2 = n$, Eq. (5.29) defines a density $\delta(\hat{\sigma}_1, \phi_{\hat{N}}{}^{t}\hat{\sigma}_2)$ on $\mathscr{T}^*Z/K$. However, $\hat{\sigma}_1$ and $\hat{\sigma}_2$ have supports in the submanifold S_n of $\mathscr{T}^*Z$ which projects onto the slice $\{ne_o\} \times \mathscr{T}^*Y$ in the product structure of $\mathscr{T}^*Z/K$ given by the diagram (10.123). Hence, the integral of $\delta(\hat{\sigma}_1, \phi_{\hat{N}}^{\ t}\hat{\sigma}_2)$ over $\mathscr{T}^*Z/K$ vanishes. To obtain a non-trivial kernel $\mathscr{K}_t$, we have to take into account the fact that we are dealing with distributional wave functions and modify $\delta(\hat{\sigma}_1, \phi_{\hat{N}}^{\ t}\hat{\sigma}_2)$ to obtain a density on $\{ne_o\} \times \mathscr{T}^*Y$ analogous to the density used in Eq. (5.13). We shall identify $\{ne_o\} \times \mathscr{T}^*Y$ with $\mathscr{T}^*Y$ and denote the resulting density on $\mathscr{T}^*Y$ by $\delta_n(\hat{\sigma}_1, \phi_{\hat{N}}^{\ t}\hat{\sigma}_2)$.

In order to define $\delta_n(\hat{\sigma}_1, \phi_{\hat{N}}^{\ t}\hat{\sigma}_2)$ we combine the arguments leading to Eqs. (5.19) and (5.29). Let z be a point in S_n, and $x = pr_2 \circ \pi_K(z)$ its projection to $\mathscr{T}^*Y$. Consider a basis of $\mathscr{T}_z^{\,c}(\mathscr{T}^*Z)$ of the form $(v, \underline{u}_1, \underline{u}_2, t)$, where

$$v \in K_z \quad \text{and} \quad \kappa(v) = 1 \tag{10.124}$$

$$\underline{w}_1 = (v, \underline{u}_1) \in \mathscr{B}_z\hat{F} \quad \text{and} \quad \underline{w}_2 = (v, \underline{u}_2) \in \mathscr{B}_z\mathscr{T}\phi_{\hat{N}}^{\ t}(\hat{F}) \tag{10.125}$$

$$i d\theta_Z(u_1^{\ j}, \bar{u}_2^{\ k}) = \hbar\delta^{jk} \quad \text{and} \quad d\theta_Z(v, t) = 1 \tag{10.126}$$

and

$$d\theta_Z(u_1^{\ j}, t) = d\theta_Z(u_2^{\ j}, t) = 0. \tag{10.127}$$

It gives rise to a basis

$$\underline{b} = (\mathscr{T}\pi_K(\underline{u}_1), \mathscr{T}\pi_K(\underline{u}_2), \mathscr{T}\pi_K(t)) \tag{10.128}$$

of $\mathscr{T}_{\pi_K(z)}(\mathscr{T}^*Z/K)$ and a basis

$$\underline{b}' = (\mathscr{T}(pr_2 \circ \pi_K)(\underline{u}_1), \mathscr{T}(pr_2 \circ \pi_K)(\underline{u}_2)) \tag{10.129}$$

of $\mathscr{T}_x(\mathscr{T}^*Y)$. Eq. (5.29) defines the value of the density $\delta(\hat{\sigma}_1, \phi_{\hat{N}}^{\ t}\hat{\sigma}_2)$ on the basis $\underline{b}$. We define $\delta_n(\hat{\sigma}_1, \phi_{\hat{N}}^{\ t}\hat{\sigma}_2)$ to have the same value on $\underline{b}'$:

$$\delta_n(\hat{\sigma}_1, \phi_{\hat{N}}^{\ t}\hat{\sigma}_2)(\underline{b}') = \delta(\hat{\sigma}_1, \phi_{\hat{N}}^{\ t}\hat{\sigma}_2)(\underline{b}). \tag{10.130}$$

One can verify by direct computation that Eq. (10.130) defines a unique density on $\mathscr{T}^*Y$.

Let $\check{q}_\alpha : V_\alpha \to R^4$ be a chart on Y and suppose that the projections of the supports of $\hat{\sigma}_1$ and $\hat{\sigma}_2$ to Y are contained in V_α. We can factorize $\hat{\sigma}_1$ and $\hat{\sigma}_2$ as follows:

$$\hat{\sigma}_i = \hat{\lambda}_i \otimes \hat{\nu}_\alpha, \tag{10.131}$$

where $\hat{\nu}_\alpha$ is defined by Eq. (10.89) and the $\hat{\lambda}_i$ are sections of $\hat{L}|S_n$ covariantly constant along $\hat{F}$. Eq. (5.41) enables us to rewrite Eq. (10.130) in the form

$$\delta_n(\hat{\sigma}_1, \phi^t_{\hat{N}}\hat{\sigma}_2)(\underline{b}') = he_o^{-1}(i/h)^2\left[\det d\theta_Z(\xi_{\hat{q}_\alpha j}, \phi^t_{\hat{N}}\xi_{\hat{q}_\alpha k})\right]^{\frac{1}{2}}(z) \times \langle\hat{\lambda}_1, \phi^t_{\hat{N}}\hat{\lambda}_2\rangle(z), \tag{10.132}$$

where the coefficient he_o^{-1} is due to the fact that the vector v in Eq. (10.124) is given by

$$v = he_o^{-1}\xi_Q(z) \tag{10.133}$$

whereas the frame field $\hat{\underline{\xi}}^\alpha$ which is used to normalize $\hat{\nu}_\alpha$ contains the vector field ξ_Q, cf. Eq. (10.88). On the other hand, we have a density $|d\theta_Y{}^4|$ on $\mathscr{T}^*Y$ which can be written in terms of the coordinates $(\underline{p}_\alpha, \underline{q}_\alpha)$ as

$$|d\theta_Y{}^4| = d^4p_\alpha d^4q_\alpha. \tag{10.134}$$

Eqs. (10.82), (10.126) and (10.129) yield

$$|d\theta_Y{}^4|(\underline{b}') = h^4. \tag{10.135}$$

Hence, comparing Eqs. (10.132), (10.134) and (10.135) we obtain

$$\delta_n(\hat{\sigma}_1,\phi_{\hat{N}}{}^t\hat{\sigma}_2)(x) = he_o{}^{-1}(i/h)^2\Big[\det d\theta_Z\Big(\xi_{\hat{q}_\alpha^j},\phi_{\hat{N}}{}^t\xi_{\hat{q}_\alpha^k}\Big)\Big]^{\frac{1}{2}}(z) \times \langle\lambda_1(z),\phi_{\hat{N}}{}^t\hat{\lambda}_2(z)\rangle\, d^4p_\alpha d^4q_\alpha. \tag{10.136}$$

We have

$$d\theta_Z\Big(\xi_{\hat{q}_\alpha^j},\phi_{\hat{N}}{}^t\xi_{\hat{q}_\alpha^k}\Big) = \omega_e\Big(\xi_{q_\alpha^j},\phi_N{}^t\xi_{q_\alpha^k}\Big)\circ pr_2\circ\pi_K \tag{10.137}$$

and

$$\langle\hat{\lambda}_1,\phi_{\hat{N}}{}^t\hat{\lambda}_2\rangle = \langle\lambda_1,\phi_N{}^t\lambda_2\rangle\circ pr_2\circ\pi_K, \tag{10.138}$$

where the sections $\hat{\lambda}_i$ and λ_i are related by Eq. (10.105). Therefore,

$$\delta_n(\hat{\sigma}_1,\phi_{\hat{N}}{}^t\hat{\sigma}_2) = he_o{}^{-1}(i/h)^2\Big[\det \omega_e\Big(\xi_{q_\alpha^j},\phi_N{}^t\xi_{q_\alpha^k}\Big)\Big]^{\frac{1}{2}} \times \langle\lambda_1,\phi_N{}^t\lambda_2\rangle\, d^4p_\alpha d^4q_\alpha. \tag{10.139}$$

The Blattner-Kostant-Sternberg kernel $\mathscr{K}_t\colon \mathscr{A}\times\mathscr{A}_t \to \mathbf{C}$ defined by

$$\mathscr{K}_t(\hat{\sigma}_1,\phi_{\hat{N}}{}^t\hat{\sigma}_2) = \int_{\mathscr{T}^*Y}\delta_n(\hat{\sigma}_1,\phi_N{}^t\sigma_2) \tag{10.140}$$

can be written

$$\mathscr{K}_t(\hat{\sigma}_1,\phi_{\hat{N}}{}^t\hat{\sigma}_2) = he_o{}^{-1}(i/h)^2\int_{\mathscr{T}^*Y}\Big\{\langle\lambda_1,\phi_N{}^t\lambda_2\rangle \times\Big[\det \omega_e\Big(\xi_{q_\alpha^j},\phi_N{}^t\xi_{q_\alpha^k}\Big)\Big]^{\frac{1}{2}}\Big\}\, d^4p_\alpha d^4p_\alpha\,. \tag{10.141}$$

Comparing Eq. (10.141) with Eq. (10.32) and taking into account Eqs. (10.31), (10.35) and (10.107), we see that

$$\mathscr{K}_t(\hat{\sigma}_1,\phi_{\hat{N}}{}^t\hat{\sigma}_2) = \mathscr{K}_t(\mathscr{U}_n\hat{\sigma}_1,\phi_N{}^t\mathscr{U}_n\hat{\sigma}_2). \tag{10.142}$$

Differentiating Eq. (10.142) with respect to t and setting t = 0, we obtain

$$\mathscr{D}_n \hat{N} = \mathscr{U}_n^{-1} (\mathscr{D} N) \mathscr{U}_n \qquad (10.143)$$

as required.

11. PAULI REPRESENTATION FOR SPIN

11.1. Classical model of spin

A classical interpretation of spin is that of an internal angular momentum. Thus, a classical state of a non-relativistic particle with spin $r > 0$ is specified by the position $\underline{q}$, the momentum $\underline{p}$, and the *spin vector* $\underline{S}$ such that

$$\underline{S}^2 = r^2. \tag{11.1}$$

Hence, the phase space under consideration is

$$X = \mathsf{R}^3 \times \mathsf{R}^3 \times S_r^{\,2}, \tag{11.2}$$

where $S_r^{\,2}$ denotes the sphere in R^3 of radius r. The Lagrange bracket for a free particle is

$$\omega_o = \sum_i dp_i \wedge dq^i - \tfrac{1}{2} r^{-2} \sum_{ijk} \varepsilon_{ijk} S_i dS_j \wedge dS_k, \tag{11.3}$$

where q^i and p_i are the usual position and momentum coordinates, respectively, and S_k are the components in R^3 of the spin vector $\underline{S}$.

The rotation group acts on X in the usual way. To the one-parameter group of rotations around the i'th axis there corresponds a vector field η_i on X given by

$$\eta_i = \sum_{jk} \varepsilon_{ijk}\left(q^j \frac{\partial}{\partial q^k} + p_j \frac{\partial}{\partial p_k} + S_j \frac{\partial}{\partial S_k}\right) . \tag{11.4}$$

The vector fields η_i are Hamiltonian with respect to the symplectic form ω_o:

$$\eta_i \lrcorner\, \omega_o = -dJ_i , \tag{11.5}$$

where

$$J_i = \sum_{jk} \varepsilon_{ijk} q^j p_k + S_i \tag{11.6}$$

is the i'th component of the total angular momentum vector

$$\underline{J} = \underline{q} \times \underline{p} + \underline{S}. \tag{11.7}$$

We consider here the dynamics of a particle with spin interacting with an external electromagnetic field. The presence of a magnetic field $\underline{B} = (B_1, B_2, B_3)$ satisfying the condition

$$\operatorname{div} \underline{B} = 0 \tag{11.8}$$

leads to the modified Lagrange bracket

$$\begin{aligned} \omega = \sum_i dp_i \wedge dq^i - \tfrac{1}{2} r^{-2} \sum_{ijk} \varepsilon_{ijk} S_i dS_j \wedge dS_k \\ + \tfrac{1}{2} e \sum_{ijk} \varepsilon_{ijk} B_i dq^j \wedge dq^k , \end{aligned} \tag{11.9}$$

where e is the electric charge of the particle. The energy in the classical state $(\underline{p}, \underline{q}, \underline{S})$ is given by

$$H(\underline{p}, \underline{q}, \underline{S}) = \underline{p}^2/2m + eV(\underline{q}) - (e/mc)\underline{S} \cdot \underline{B}(\underline{q}) . \tag{11.10}$$

One could add to H a term corresponding to the spin-orbit coupling but this will not be done here. The Hamiltonian vector field ξ_H, defined via the symplectic form ω, is

$$\xi_H = m^{-1} \sum_i p_i \frac{\partial}{\partial q^i} - (e/mc) \sum \varepsilon_{ijk} B_j S_k \frac{\partial}{\partial S_i} \tag{11.11}$$
$$+ e \sum_i \left\{ \frac{\partial V}{\partial q^i} + m^{-1} \sum_{jk} \varepsilon_{ijk} p_j B_k - (mc)^{-1} \sum_j S_j \frac{\partial B_j}{\partial q^i} \right\} \frac{\partial}{\partial p_i}.$$

The local one-parameter group $\phi_H{}^t$ of canonical transformations of (X,ω) generated by ξ_H satisfies the following equations:

$$\frac{d}{dt}(\underline{q} \circ \phi_H{}^t) = m^{-1} \underline{p} \circ \phi_H{}^t \tag{11.12}$$

$$\frac{d}{dt}(\underline{p} \circ \phi_H{}^t) = - \{\text{grad } V + m^{-1} \underline{p} \times \underline{B} - (mc)^{-1} \text{grad}(\underline{S} \cdot \underline{B})\} \circ \phi_H{}^t \tag{11.13}$$

$$\frac{d}{dt}(\underline{S} \circ \phi_H{}^t) = (e/mc)(\underline{S} \times \underline{B}) \circ \phi_H{}^t, \tag{11.14}$$

where grad denotes the gradient with respect to the position variables $\underline{q}$.

11.2. Representation space

Let V_+ and V_- be open sets in X defined by

$$V_\pm = \{x \in X | r \pm S_3(x) \neq 0\}. \tag{11.15}$$

We denote by z_+ and z_- the complex functions on V_+ and V_-, respectively, defined by

$$z_\pm = \frac{S_1 \mp iS_2}{r \pm S_3}. \tag{11.16}$$

In $V_+ \cap V_-$ we have

$$z_+ z_- = 1, \tag{11.17}$$

and the functions z_+ and z_- define a complex structure on each sphere of constant position and momenta. Solving Eq. (11.16) for $\underline{S}$ in terms of z_+ and z_-, we obtain

$$S_1 = r(z_\pm+\bar{z}_\pm)(1+z_\pm\bar{z}_\pm)^{-1} \tag{11.18}$$

$$S_2 = \pm ir(z_\pm-\bar{z}_\pm)(1+z_\pm\bar{z}_\pm)^{-1} \tag{11.19}$$

$$S_3 = \pm r(1-z_\pm\bar{z}_\pm)(1+z_\pm\bar{z}_\pm)^{-1}. \tag{11.20}$$

This enables us to express ω in terms of the complex coordinates $z_\pm$ as follows:

$$\omega|V_\pm = \sum_i dp_i \wedge dq^i + \tfrac{1}{2}e \sum_{ijk} \varepsilon_{ijk} B_i dq^j \wedge dq^k - 2ir(1+z_\pm\bar{z}_\pm)^{-2}\, d\bar{z}_\pm \wedge dz_\pm. \tag{11.21}$$

Consider the linear frame fields $\underline{\xi}_\pm$ on $V_\pm$ defined by

$$\underline{\xi}_\pm = (\xi_{q^1}, \xi_{q^2}, \xi_{q^3}, \xi_{z_\pm}). \tag{11.22}$$

For each $x \in V_+ \cap V_-$,

$$\underline{\xi}_-(x) = \underline{\xi}_+(x) \begin{bmatrix} 1 & 0 & 0 & 0 \\ 0 & 1 & 0 & 0 \\ 0 & 0 & 1 & 0 \\ 0 & 0 & 0 & -z_+^{-2}(x) \end{bmatrix}. \tag{11.23}$$

Hence, $\underline{\xi}_+$ and $\underline{\xi}_-$ span a complex distribution F on X. Eq. (11.21) yields

$$\xi_{q^i} = -\frac{\partial}{\partial p_i} \quad \text{and} \quad \xi_{z_\pm} = (2ir)^{-1}(1+z_\pm\bar{z}_\pm)^2 \frac{\partial}{\partial \bar{z}_\pm} \tag{11.24}$$

which implies that the frame fields $\underline{\xi}_\pm$ and the distribution F are independent of the choice of the magnetic field $\underline{B}$ and the value of the charge e. For each $\underline{B}$ and e, the distribution F is a polarization of (X,ω). The space X/D of integral manifolds of $D = F \cap \bar{F} \cap \mathscr{T}X$ can be identified with $R^3 \times S_r^2$, and the canonical projection $\pi_D\colon X \to X/D$ corresponds to the projection of $R^3 \times (R^3 \times S_r^2)$ onto the second factor $R^3 \times S_r^2$. Similarly, the space X/E of integral manifolds of $E = (F + \overline{F}) \cap \mathcal{T}X$ can be identified with R^3, and the canonical projection $\pi_{DE}: X/D \to X/E$ corresponds to the projection onto the first factor of $R^3 \times S_r^2$. Moreover,

$$i\omega(\xi_{z_\pm}, \overline{\xi}_{z_\pm}) = (1+z_\pm\overline{z}_\pm)^2/2r > 0 \tag{11.25}$$

so that F is a complete strongly admissible positive polarization of (X,ω).

Let L be a prequantization line bundle of (X,ω). The prequantization condition (3.20) is satisfied if and only if the de Rham cohomology class $[-h^{-1}\omega]$ of $-h^{-1}\omega$ is integral. Since the magnetic field $\underline{B}$ on R^3 is globally defined, Eq. (11.8) implies that there exists a vector potential for $\underline{B}$ so that the last term on the right hand side of Eq. (11.9) is an exact form. Hence,

$$[-h^{-1}\omega] = [\tfrac{1}{2}h^{-1}r^{-2}\sum_{ijk}\varepsilon_{ijk}S_i dS_j \wedge dS_k]. \tag{11.26}$$

Integrating the form on the right hand side of Eq. (11.9) over the sphere $\{\underline{0}\} \times \{\underline{0}\} \times S_r^2 \subseteq X$ we obtain $4\pi r h^{-1}$ which must be an integer if $[-h^{-1}\omega]$ is integral. Hence, the prequantization condition is satisfied if and only if

$$2r\hbar^{-1} \in Z. \tag{11.27}$$

If we write

$$r = \hbar s \tag{11.28}$$

then (11.27) implies that $2s$ must be an integer,

$$2s \in Z. \tag{11.29}$$

Thus, the prequantization condition leads to the quantization of spin. The integer $2s$ is the *Chern class* of the bundle L.

The restrictions of ω to the open sets V_+ and V_- are exact:

$$\omega|V_\pm = d\theta_\pm, \tag{11.30}$$

where

$$\theta_\pm = \sum_j (p_j + eA_j(\underline{q}))dq^j - 2ir(1+z_\pm\bar{z}_\pm)^{-1}\bar{z}_\pm dz_\pm \tag{11.31}$$

and $\underline{A}$ is a vector potential for $\underline{B}$ satisfying

$$\underline{B} = \text{curl } \underline{A}. \tag{11.32}$$

Let $\lambda_\pm$ be local sections of $L|V_\pm$ such that

$$\nabla\lambda_\pm = -i\hbar^{-1}\,\theta_\pm \otimes \lambda_\pm. \tag{11.33}$$

In $V_+ \cap V_-$ we have

$$-i\hbar^{-1}(\theta_+ - \theta_-) = d \log z_-^{2r/\hbar}. \tag{11.34}$$

Taking into account Eq. (11.28) we see that the transition function

$$z_-^{2r/\hbar} = z_-^{2s} \tag{11.35}$$

is globally defined and single-valued on $V_+ \cap V_-$ since $2s \in \mathbb{Z}$. Hence, we may assume that λ_+ and λ_- are normalized so that

$$\lambda_+ = z_-^{2s}\lambda_-. \tag{11.36}$$

The connection invariant Hermitian metric on L satisfies the equation

$$d\langle\lambda_\pm,\lambda_\pm\rangle = -i\hbar\langle\lambda_\pm,\lambda_\pm\rangle(\theta_\pm - \bar{\theta}_\pm), \tag{11.37}$$

whence

$$\langle\lambda_\pm,\lambda_\pm\rangle = (1+z_\pm\bar{z}_\pm)^{-2s}. \tag{11.38}$$

The sections λ_+ and λ_- are covariantly constant along F and consequently each covariantly constant section λ of L can be represented as follows:

$$\lambda|V_{\pm} = \psi_{\pm}(\underset{\sim}{q}, z_{\pm})\lambda_{\pm}, \tag{11.39}$$

where $\psi_{\pm}$ is a holomorphic function of $z_{\pm}$. Taking into account Eqa. (11.36) and (11.17), we obtain the relation

$$\psi_{+}(\underset{\sim}{q}, z_{+}) = z_{+}{}^{2s}\psi_{-}(\underset{\sim}{q}, z_{+}{}^{-1}) \tag{11.40}$$

on $V_{+} \cap V_{-}$. Since both ψ_{+} and ψ_{-} are holomorphic functions of the second argument, Eq. (11.40) implies that they are polynomials of degree at most $2s$ in this variable. For each integer m such that $0 \leq m \leq 2s$, we denote by λ_m the section of L defined by

$$\lambda_m|V_{+} = c_m z_{+}{}^{m}\lambda_{+} \quad \text{and} \quad \lambda_m|V_{-} = c_m z_{-}{}^{2s-m}\lambda_{-}, \tag{11.41}$$

where the normalization constants c_m are chosen so that

$$\int_{S_r^2} \langle\lambda_m, \lambda_m\rangle\omega = 1 \tag{11.42}$$

for each sphere S_r^2 in X corresponding to constant values of the position and the momenta variables. The sections λ_m are covariantly constant along F, and each covariantly constant section λ of L is of the form

$$\lambda = \sum_{m} \psi_m(\underset{\sim}{q})\lambda_m, \tag{11.43}$$

where the $\psi_m(\underset{\sim}{q})$ are functions of the position variables only.

According to the general theory of geometric quantization the representation space should consist of those sections of $L \otimes \sqrt{\wedge^4 F}$ which are covariantly constant along F. Since the second Chern class of $\sqrt{\wedge^4 F}$ is 2 the Chern class of

$\sqrt{\wedge^4 F}$ is 1. Therefore, $L \otimes \sqrt{\wedge^4 F}$ has Chern class 2s+1 and the space of holomorphic sections of $L \otimes \sqrt{\wedge^4 F}$ restricted to any sphere corresponding to constant values of the position and momenta variables has dimension 2s+2, cf. Gunning (1966), p. 111. Hence, the choice of the space of covariantly constant sections of $L \otimes \sqrt{\wedge^4 F}$ for the representation space leads to a wrong dimension for the space of spin states. In order to circumvent this difficulty we choose the space of square integrable sections of $L \otimes \sqrt{\wedge^3 D^{\mathbf{C}}}$ which are covariantly constant along F for the representation space $\mathcal{H}$. This choice is justified by the agreement between the quantum theory it engenders and the Pauli theory of spin. It requires certain modifications of the procedures developed in Chapters 4 and 5, the theoretical significance of which has yet to be studied.

The distribution $D^{\mathbf{C}}$ is globally spanned by the frame field

$$\underline{\xi} = \left(\frac{\partial}{\partial p_1}, \frac{\partial}{\partial p_2}, \frac{\partial}{\partial p_3}\right). \tag{11.44}$$

We denote by $\underline{\tilde{\xi}}$ a metalinear frame field for $D^{\mathbf{C}}$ which projects onto $\underline{\xi}$ and by $\nu_{\underline{\tilde{\xi}}}$ the section of $\sqrt{\wedge^3 D^{\mathbf{C}}}$ defined by

$$\nu_{\underline{\tilde{\xi}}}^{\#} \circ \underline{\tilde{\xi}} = 1. \tag{11.45}$$

The section $\nu_{\underline{\tilde{\xi}}}$ is covariantly constant along F and thus every covariantly constant section ν of $\sqrt{\wedge^3 D^{\mathbf{C}}}$ is of the form

$$\nu = \psi(\underline{q})\nu_{\underline{\tilde{\xi}}}, \tag{11.46}$$

where $\psi(\underline{q})$ is a complex-valued function of the position variables only. The function ψ does not depend on the complex variables $z_{\pm}$ since every holomorphic function on a sphere is constant.

The representation space $\mathscr{H}$ consists of covariantly constant sections of $L \otimes \sqrt{\wedge^3 D^{\mathbf{C}}}$. Taking into account Eqs. (11.39) and (11.42), we see that every section σ in $\mathscr{H}$ is of the form

$$\sigma = \sum_m \psi_m(\underline{q})\lambda_m \otimes \nu_{\underline{\tilde{\xi}}}. \tag{11.47}$$

The scalar product on $\mathscr{H}$ is given as follows: for any two sections σ and σ' of the form (11.47),

$$(\sigma|\sigma') = \sum_m \int_{\mathbf{R}^3} \psi_m(\underline{q})\overline{\psi}_m{}'(\underline{q})d^3q. \tag{11.48}$$

Thus, the mapping associating to each $\lambda \otimes \nu_{\underline{\tilde{\xi}}} \in \mathscr{H}$ the function $\underline{\psi}$: $\mathbf{R}^3 \to \mathbf{C}^{2s+1}$ such that

$$\underline{\psi}(\underline{q}) = [\psi_o(\underline{q}),\ldots,\psi_{2s}(\underline{q})], \tag{11.49}$$

where $\psi_o(\underline{q}),\ldots,\psi_{2s}(\underline{q})$ are given by Eq. (11.47), is a unitary isomorphism of $\mathscr{H}$ onto the space of square integrable $\mathbf{C}^{2s+1}$-valued functions on $\mathbf{R}^3$.

11.3. Quantization

The canonical transformations generated by the components of the position, the momentum, the spin and the angular momentum vectors preserve the polarization F. Hence they preserve $D^{\mathbf{C}} = F \cap \overline{F}$, and we can use the results of Sec. 6.2 to evaluate the quantum operators corresponding to these dynamical variables.

Since the position functions q_i are constant along F, Eqs. (6.26) and (11.47) yield

$$\mathscr{Q}q_i\Big[\sum_m \psi_m(\underline{q})\lambda_m \otimes \nu_{\underline{\tilde{\xi}}}\Big] = \sum_m q_i\psi_m(\underline{q})\lambda_m \otimes \nu_{\underline{\tilde{\xi}}}. \tag{11.50}$$

The Hamiltonian vector field of p_j is

$$\xi_{p_j} = \frac{\partial}{\partial q^j} - e \sum_{ik} \varepsilon_{ijk} B_i \frac{\partial}{\partial p_k} . \tag{11.51}$$

Taking into account Eqs. (6.24), (11.31), (11.33), (11.39), (11.47) and (11.51), we obtain

$$\mathcal{Q}p_j\left[\sum_m \psi_m(\underline{q})\lambda_m \otimes \nu_{\tilde{\underline{\xi}}}\right] = \sum_m \left\{\left[-i\hbar \frac{\partial}{\partial q^j} - eA_j(\underline{q})\right]\psi_m(\underline{q})\right\} \lambda_m \otimes \nu_{\tilde{\underline{\xi}}} . \tag{11.52}$$

The Hamiltonian vector fields ξ_{S_1}, ξ_{S_2} and ξ_{S_3} can be expressed with the help of Eqs. (11.18), (11.19), (11.20) and (11.21) as follows:

$$\xi_{S_1}\Big|V_\pm = -\frac{i}{2}\left[(z_\pm{}^2-1)\frac{\partial}{\partial z_\pm} - (\bar{z}_\pm{}^2-1)\frac{\partial}{\partial \bar{z}_\pm}\right] \tag{11.53}$$

$$\xi_{S_2}\Big|V_\pm = \pm\frac{1}{2}\left[(z_\pm{}^2+1)\frac{\partial}{\partial z_\pm} + (\bar{z}_\pm{}^2+1)\frac{\partial}{\partial \bar{z}_\pm}\right] \tag{11.54}$$

$$\xi_{S_3}\Big|V_\pm = \pm i\left[\bar{z}_\pm \frac{\partial}{\partial \bar{z}_\pm} - z_\pm \frac{\partial}{\partial z_\pm}\right]. \tag{11.55}$$

These equations, together with Eqs. (6.24), (11.31) and (11.33), yield

$$\mathcal{Q}S_1[\lambda_\pm \otimes \nu_{\tilde{\underline{\xi}}}] = rz_\pm\lambda_\pm \otimes \nu_{\tilde{\underline{\xi}}} \tag{11.56}$$

$$\mathcal{Q}S_2[\lambda_\pm \otimes \nu_{\tilde{\underline{\xi}}}] = irz_\pm\lambda_\pm \otimes \nu_{\tilde{\underline{\xi}}} \tag{11.57}$$

$$\mathcal{Q}S_3[\lambda_\pm \otimes \nu_{\tilde{\underline{\xi}}}] = \pm r\lambda_\pm \otimes \nu_{\tilde{\underline{\xi}}} . \tag{11.58}$$

Hence,

$$\mathcal{Q}S_1[\psi_\pm\lambda_\pm\otimes\nu_{\tilde{\underline{\xi}}}] = \left\{\left[-\tfrac{1}{2}\hbar(z_\pm{}^2-1)\frac{\partial}{\partial z_\pm} + rz_\pm\right]\psi_\pm\right\}\lambda_\pm \otimes \nu_{\tilde{\underline{\xi}}} \tag{11.59}$$

$$\mathcal{Q}S_2[\psi_\pm\lambda_\pm\otimes\nu_{\tilde{\underline{\xi}}}] = \left\{\left[\mp\tfrac{1}{2}i\hbar(z_\pm{}^2+1)\frac{\partial}{\partial z_\pm} + irz_\pm\right]\psi_\pm\right\}\lambda_\pm \otimes \nu_{\tilde{\underline{\xi}}} \tag{11.60}$$

$$\mathcal{Q}S_3[\psi_\pm\lambda_\pm\otimes\nu_{\tilde{\underline{\xi}}}] = \left\{\pm\left[-\hbar z_\pm \frac{\partial}{\partial z_\pm} + r\right]\psi_\pm\right\}\lambda_\pm \otimes \nu_{\tilde{\underline{\xi}}} . \tag{11.61}$$

Applying these results to the basis vectors $\lambda_m \otimes \nu_{\underline{\tilde{\xi}}}$, where λ_m is defined by Eq. (11.41), and taking into account the fact that $r = s\hbar$, we obtain

$$\mathscr{Q}s_1[\lambda_m \otimes \nu_{\underline{\tilde{\xi}}}] = \frac{1}{2}\hbar\left[(2s-m)c_m c_{m+1}^{-1}\lambda_{m+1} + mc_m c_{m-1}^{-1}\lambda_{m-1}\right]\otimes \nu_{\underline{\tilde{\xi}}} \quad (11.62)$$

$$\mathscr{Q}s_2[\lambda_m \otimes \nu_{\underline{\tilde{\xi}}}] = \frac{1}{2}i\hbar\left[(2s-m)c_m c_{m+1}^{-1}\lambda_{m+1} - mc_m c_{m-1}^{-1}\lambda_{m-1}\right]\otimes \nu_{\underline{\tilde{\xi}}} \quad (11.63)$$

$$\mathscr{Q}s_3[\lambda_m \otimes \nu_{\underline{\tilde{\xi}}}] = \hbar(s-m)\lambda_m \otimes \nu_{\underline{\tilde{\xi}}}. \quad (11.64)$$

For each $i = 1,2,3$, let $\underline{D}_i$ be the $(2s+1) \times (2s+1)$ matrix with entries $d_{i;m,n}$ such that, for each $m \in \{0,1,\ldots,2s\}$,

$$d_{1;m,m+1} = \frac{1}{2}\hbar(2s-m)c_m c_{m+1}^{-1} \quad (11.65)$$

$$d_{1;m,m-1} = \frac{1}{2}\hbar m c_m c_{m+1}^{-1} \quad (11.66)$$

$$d_{2;m,m+1} = \frac{1}{2}i\hbar(2s-m)c_m c_{m+1}^{-1} \quad (11.67)$$

$$d_{2;m,m-1} = -\frac{1}{2}i\hbar m c_m c_{m-1}^{-1} \quad (11.68)$$

$$d_{3;m,m} = \hbar(s-m) \quad (11.69)$$

and all other entries vanish. Then, for each section in $\mathscr{A}$ of the form (11.47), we have

$$\mathscr{Q}s_i\left[\sum_m \psi_m(\underline{q})\lambda_m \otimes \nu_{\underline{\tilde{\xi}}}\right] = \sum_{mn} d_{i;m,n}\psi_m(\underline{q})\lambda_n \otimes \nu_{\underline{\tilde{\xi}}} \quad (11.70)$$

for each $i = 1,2,3$. The quantization of the total angular momentum vector $\underline{J}$ is obtained in an analogous manner, yielding

$$\mathscr{Q}J_i\Big[\sum_m \psi_m(\underline{q})\lambda_m \otimes \nu_{\tilde{\underline{\xi}}}\Big]$$

$$= \sum_m \Big\{\sum_{jk} \varepsilon_{ijk} q^j \Big[-i\hbar\frac{\partial}{\partial q^k} - eA_k(\underline{q})\Big]\psi_m(\underline{q})\Big\}\lambda_m \otimes \nu_{\tilde{\underline{\xi}}} \qquad (11.71)$$

$$+ \sum_{mn} d_{i;m,n}\psi_m(\underline{q})\lambda_n \otimes \nu_{\tilde{\underline{\xi}}}.$$

It remains to obtain the quantum operator $\mathscr{Q}H$ corresponding to the energy H given by Eq. (11.10). The one-parameter group $\phi_H{}^t$ of canonical transformations of (X,ω) generated by H does not preserve $D^{\mathbf{c}}$, and consequently we cannot apply the technique developed in Sec. 6.2. Since we are using sections of $L \otimes \sqrt{\wedge^3 D^{\mathbf{c}}}$ rather than $L \otimes \sqrt{\wedge^4 F}$ to define the representation space $\mathscr{H}$, we cannot quantize directly via the Blattner-Kostant-Sternberg kernels defined in Chapter 5 without first modifying them in order to be applicable in the present situation. To do so we assume that, for sufficiently small positive t, the distribution

$$G_t = D + \mathscr{T}\phi_H{}^t D \qquad (11.72)$$

has dimension 6, is involutive and, for each integral manifold M of G_t, the restriction of ω to M is a symplectic form on M. These assumptions are satisfied in the absence of the magnetic field and we may expect that they will hold for sufficiently weak magnetic fields. The restrictions of $D^{\mathbf{c}}$ and $\mathscr{T}\phi_H{}^t D^{\mathbf{c}}$ to the integral manifolds M of G_t are transverse pairs of real polarizations of $(M,\omega|M)$. Hence, if each $(M,\omega|M)$ is given a metaplectic structure, we can associate to each pair of sections σ_M and σ_{tM} of $L \otimes \sqrt{\wedge^3 D^{\mathbf{c}}}|M$ and $L \otimes \sqrt{\wedge^3 \mathscr{T}\phi_H{}^t(D^{\mathbf{c}})}|M$, respectively, the function $\langle\sigma_M,\sigma_{tM}\rangle$ on M given by Eq. (5.11). Thus, if σ and σ_t are sections of

$L \otimes \sqrt{\wedge^3 D^{\mathbb{C}}}$ and $L \otimes \sqrt{\wedge^3 \mathscr{T}\phi_H{}^t(D^{\mathbb{C}})}$, respectively, we can associate to them a function $\langle\sigma,\sigma_t\rangle$ on X such that, for each M,

$$\langle\sigma,\sigma_t\rangle|M = \langle\sigma_M,\sigma_{tM}\rangle, \tag{11.73}$$

where

$$\sigma_M = \sigma|M \quad \text{and} \quad \sigma_{tM} = \sigma_t|M. \tag{11.74}$$

The generalized Blattner-Kostant-Sternberg kernel $\mathscr{K}: \mathscr{A} \times \mathscr{A}_t \to \mathbb{C}$ is given by

$$\mathscr{K}(\sigma,\sigma_t) = \int_X \langle\sigma,\sigma_t\rangle|\omega^4|. \tag{11.75}$$

As before, let $\mathscr{U}_t: \mathscr{A}_t \to \mathscr{A}$ be the linear map defined by

$$\mathscr{K}(\sigma,\sigma_t) = (\sigma|\mathscr{U}_t\sigma_t), \tag{11.76}$$

and let $\Phi_t: \mathscr{A} \to \mathscr{A}$ be given by

$$\Phi_t(\sigma) = \mathscr{U}_t(\phi_H{}^t\sigma). \tag{11.77}$$

The quantum operator $\mathscr{Q}H$ corresponding to the classical energy H is then

$$\mathscr{Q}H = i\hbar \frac{d}{dt} \Phi_t\Big|_{t=0}. \tag{11.78}$$

According to Eqs. (11.76) and (11.77) we can write

$$\begin{aligned}&\Big(\sum_m \psi_m'(\underline{q})\lambda_m \otimes \nu_{\tilde{\underline{\xi}}} \,\Big|\, \Phi_t\Big[\sum_n \psi_n(\underline{q})\lambda_n \otimes \nu_{\tilde{\underline{\xi}}}\Big]\Big) \\ &= \mathscr{K}_t\Big(\sum_m \psi_m'(\underline{q})\lambda_m \otimes \nu_{\tilde{\underline{\xi}}},\ \sum_n \psi_n(q\circ\phi_H{}^{-t})\phi_H{}^t\lambda_n \otimes \phi_H{}^t\nu_{\tilde{\underline{\xi}}}\Big),\end{aligned} \tag{11.79}$$

where $\phi_H{}^t\lambda_m$ is defined by Eq. (3.32) and $\phi_H{}^t\nu_{\tilde{\underline{\xi}}}$ by Eq. (6.2). Given $2s+1$ functions $\psi_m(\underline{q})$, let $\psi_{tm}(\underline{q})$ be defined by

$$\Phi_t\Big[\sum_m \psi_m(\underline{q})\lambda_m \otimes \nu_{\tilde{\underline{\xi}}}\Big] = \sum_m \psi_{tm}(\underline{q})\lambda_m \otimes \nu_{\tilde{\underline{\xi}}}. \tag{11.80}$$

Substituting this expression into Eq. (11.79) and taking into account Eqs. (5.11), (11.73), (11.75) and (11.48), we obtain

$$\sum_m \int_{R^3} \psi_m'(\underline{q})\overline{\psi}_{tm}(\underline{q})\,d^3q = (i/\hbar)^{3/2}\int_X \Big\{\Big[\det\,\omega\Big(\xi_{q^j}, \phi_H{}^t \xi_{q^k}\Big)\Big]^{\frac{1}{2}}$$
$$\times \sum_{mn} \langle\lambda_m, \phi_H{}^t\lambda_n\rangle\; \psi_m'(\underline{q})\overline{\psi}_n(\underline{q}\circ\phi_H{}^{-t})\Big\}|\omega^4|. \tag{11.81}$$

It follows from Eq. (11.21) that

$$|\omega^4|\,\Big|V_+ = d^3p\,d^3q\,|\omega\Big|S_r{}^2|. \tag{11.82}$$

Hence

$$\psi_{tm}(\underline{q}) = (i\hbar)^{-3/2}\int_{R^3\times S_r{}^2}\Big\{\Big[\det\omega\Big(\xi_{q^j}, \phi_H{}^t\xi_{q^k}\Big)\Big]^{\frac{1}{2}}$$
$$\times \sum_n \langle\phi_H{}^t\lambda_n, \lambda_m\rangle\psi_n(\underline{q}\circ\phi_H{}^{-t})\Big\}\; d^3p\,|\omega\Big|S_r{}^2|. \tag{11.83}$$

Following the arguments leading to Eq. (6.48) and bearing in mind the results of the quantization of the spin vector $\underline{S}$, we obtain

$$\langle\phi_H{}^t\lambda_n, \lambda_m\rangle = \exp\Big\{i\hbar^{-1}\int_0^t [(\underline{p}^2/2m) + (e/m)\underline{p}\cdot\underline{A}(\underline{q}) - eV(\underline{q})]\circ\phi_H{}^{-u}du\Big\}$$
$$\times \sum_k\Big\{\exp\Big[(ie/mc\hbar)\sum_j \int_0^t B_j(\underline{q}\circ\phi_H{}^{-u})\,du\;\underline{D}_j\Big]\Big\}_{nk}\langle\lambda_k, \lambda_m\rangle, \tag{11.84}$$

where the matrices $\underline{D}_j$ are given by Eqs. (11.65) through (11.69). The exponential function in the second line is defined in terms of matrix multiplication, i.e., for any square matrix $\underline{C}$,

$$\exp\,\underline{C} = \sum_{n=0}^{\infty}(n!)^{-1}\underline{C}^n. \tag{11.85}$$

Taking into account Eqs. (11.40) and (11.41) we obtain

$$\psi_{tm}(\underline{q}) = (i\hbar)^{-3/2}\sum_n \int_{R^3}\left\{\exp\left[(ie/mc\hbar)\sum_j \int_0^t B_j(\underline{q}\circ\phi_H{}^{-u})du\ \underline{D}_j\right]\right\}_{nm}$$

$$\times\ \psi_n(\underline{q}\circ\phi_H{}^{-t})\left[\det\ \omega(\xi_{q^j},\phi_H{}^t\xi_{q^k})\right]^{\frac{1}{2}} \tag{11.86}$$

$$\times\ \exp\left\{-i\hbar^{-1}\int_0^t [(p^2/2m)+(e/m)\underline{p}\cdot\underline{A}(\underline{q}) - eV(\underline{q})]\circ\phi_H{}^{-u}\right\}d^3p.$$

We can evaluate the derivative of $\psi_{tm}(\underline{q})$ with respect to t at $t = 0$ in the same way as in Sec. 7.1 and Sec. 10.1. Using approximations analogous to those leading to Eqs. (7.38) and (10.56), we obtain

$$\lim_{t\to 0+} \frac{d}{dt}\psi_{tm}(\underline{q})$$

$$= i\hbar^{-1}(e/mc)\sum_{nj} B_j d_{j;n,m}\psi_n(\underline{q}) - i\hbar^{-1}eV(\underline{q})\psi_m(\underline{q}) \tag{11.87}$$

$$+\ (i\hbar/2m)\sum_{jk}\left(\frac{\partial}{\partial q^j} - ie\hbar^{-1}A_j\right)\left(\frac{\partial}{\partial q^k} - ie\hbar^{-1}A_k\right)\psi_m(\underline{q}).$$

Eqs. (11.78), (11.80) and (11.87) yields

$$\mathscr{D}H\left[\sum_m \psi_m(\underline{q})\lambda_m\otimes\nu_{\underline{\tilde\xi}}\right] = (-e/mc)\sum_{mnj} B_j d_{j;n,m}\psi_n(\underline{q})\lambda_m\otimes\nu_{\underline{\tilde\xi}} \tag{11.88}$$

$$+\ \sum_m\left\{\left[-\frac{\hbar^2}{2m}\sum_{jk}\left(\frac{\partial}{\partial q^j} - ie\hbar^{-1}A_j\right)\left(\frac{\partial}{\partial q^j} - ie\hbar^{-1}A_k\right)+V(\underline{q})\right]\psi_m(\underline{q})\right\}\lambda_m\otimes\nu_{\underline{\tilde\xi}}.$$

Taking into account Eq. (11.70) we can rewrite Eq. (11.88) in the form

$$\mathscr{D}H\left[\sum_m \psi_m\lambda_m\otimes\nu_{\underline{\tilde{\xi}}}\right] = (-e/mc)\sum_j B_j\mathscr{D}S_j\left[\sum_m \psi_m\lambda_m\otimes\nu_{\underline{\tilde{\xi}}}\right]$$
$$+ \sum_m\left\{\left[-\frac{\hbar^2}{2m}\sum_{jk}\left(\frac{\partial}{\partial q^j} - ie\hbar^{-1}A_j\right)\left(\frac{\partial}{\partial q^k} - ie\hbar^{-1}A_k\right) + eV\right]\psi_m\right\}\lambda_m\otimes\nu_{\underline{\tilde{\xi}}}. \tag{11.89}$$

For $s = \frac{1}{2}$, Eq. (11.89) corresponds to the Pauli equation.

BIBLIOGRAPHY

Abraham, R. and Marsden, J.E. (1978). *Foundations of mechanics*, 2nd ed. Addision-Wesley, Benjamin/Cummings, Reading, Mass.

Arens, R. (1971). Classical relativistic particles. *Comm. Math. Phys.* Vol. 21, pp. 139-149.

Arnold, V. I. (1967). Characteristic class entering in quantization conditions. *Funct. Anal. Appl.* Vol. 1, pp. 1-13.

Auslander, L. and Kostant, B. (1971). Polarizations and unitary representations of solvable Lie groups. *Invent. Math.* Vol. 14, pp. 255-354.

Bargmann, V. (1961). On a Hilbert space of analytic functions and an associated integral transform. *Comm. Pure and Appl. Math.* Vol. 14, pp. 187-214.

Bargmann, V. (1962). Remarks on a Hilbert space of analytic functions. *Proc. Nat. Acad. Sci. U.S.A.* Vol. 48, pp. 199-204.

Baxter, J. "Non-relativistic quantum theory of spin," in preparation.

Ben-Abraham, S.I. and Lonke, A. (1973). Quantization of a general dynamical system. *J. Math. Phys.* Vol. 14, pp. 1935-1937.

Bez, H. (1977). Quantization of Galilean systems. *Int. J. Theor. Phys.* Vol. 16, pp. 7-12.

Blackman, J. (1976). Geometric quantization and Feynman path integral. Preliminary report. Department of Mathematics, Syracuse University.

Blattner, R.J. (1973). Quantization and representation theory, in *Harmonic analysis on homogeneous spaces*. Edited by E.T. Taam. *Proc. Sym. Pure Math.* Vol. 26, pp. 147-165. A.M.S., Providence, R.I.

Blattner, R.J. (1975a). Intertwining operators and the half-density pairing. *Lecture Notes in Mathematics*. Vol. 466, pp. 1-12. Springer, New York.

Blattner, R.J. (1975b). Pairing of half-form spaces, in *Géométrie symplectique et physique mathématique*. Coll. Int. C.N.R.S. No. 237, pp. 175-186. C.N.R.S., Paris.

Blattner, R.J. (1977). The metalinear geometry of non-real polarizations. *Lecture Notes in Mathematics*. Vol. 570, pp. 11-45, Springer, New York.

Burdet, G. and Perrin, M. (1978). Weyl quantization and metaplectic representation. *Lecture Notes in Physics*. Vol. 79, pp. 515-517. Springer, New York.

Campbell, P. (1978). Projective geometry for polarizations. *Conference on differential-geometric methods in physics*. Clausthal.

Carey, A.L. and Hannabus, K.C. (1978). Twistors and geometric quantization theory. *Rep. Math. Phys*. Vol. 13, pp. 199-231.

Cressman, R. (1977). Quantization and group representations. *Can. J. Math*. Vol. 29, pp. 1264-1276.

Czyż, J. (1978a). On a modification of the Kostant-Souriau geometric quantization procedure. *Bull. Acad. Polon. Sci., Sér. Sci. math., astr. et phys*. Vol. 26, pp. 129-138.

Czyż, J. (1978b). On some approach to geometric quantization. *Lecture Notes in Mathematics*. Vol. 676, pp. 315-328. Springer, New York.

Czyż, J. (1978c). On geometric quantization and its connection with the Maslov theory. Preprint. Institute of Mathematics, Technical University, Warsaw.

DeWitt, B.S. (1957). Dynamical theory in curved spaces, I. A review of the classical and quantum action principles. *Rev. Mod. Phys*. Vol. 29, pp. 377-397.

Dowker, J.S. (1975). Covariant Schrödinger equations, in *Functional integration and its applications*, pp. 34-52. Edited by A.M. Arthurs. Clarendon Press, Oxford.

Duval, Ch. (1976). The general relativistic Dirac-Pauli particle: an underlying classical model. *Ann. Inst. Henri Poincaré*, Sec. A, Vol. 25, pp. 345-362.

Elhadad, J. (1974). Sur l'interprétation en géométrie symplectique des états quantiques de l'atome d'hydrogène. *Symposia mathematica*, Vol. 14, pp. 259-291. Academic Press, London.

Elhadad, J. (1977). Quantification du flot geodesique de la sphere S^n. Preprint. Centre de Physique Théorique, C.N.R.S., Marseille.

Fermi, E. (1962). Notes on Quantum Mechanics, Phenix Science Series, The University of Chicago Press, Chicago.

Fock, V. (1928). Verallgemeinerung und Lösung det Diracschen statistischen Gleichung. *Z. F. Physik*, Vol. 49, pp. 339-357.

Forger, M. and Hess, H. (1978). Universal metaplectic structures and geometric quantization. Preprint. Inst. of Theor. Phys. Freie Universität Berlin.

Gawędzki, K. (1975). On extensions of Kostant quantization procedure, in *Géométrie symplectique et physique mathématique*. Coll. Int. C.N.R.S. No. 237, pp. 211-216, C.N.R.S. Paris.

Gawędzki, K. (1977). Geometric quantization and Feynman path integrals for spin. *Lecture Notes in Mathematics*. Vol. 570, pp. 67-71. Springer, New York.

Gotay, M.J. and Isenberg, J. (1978). Kostant-Souriau quantization of Robertson-Walker cosmologies with a scalar field. *Lecture Notes in Physics*. Vol. 94, pp. 293-295.

Guillemin, V. (1975). Symplectic spinors and partial differential equations, in *Géométrie symplectique et physique mathématique*. Coll. Int. C.N.R.S. No. 237, pp. 217-252. C.N.R.S. Paris.

Guillemin, V. and Sternberg, S. (1977). *Geometric asymptotics*. A.M.S., Providence, R.I.

Günther, C. (1977). Prequantum bundles and projective Hilbert geometries. *Int. J. Theor. Phys*. Vol. 16, pp. 447-464.

Hess, H. (1978a). On a geometric quantization scheme generalizing those of Kostant-Souriau and Czyż. *Conference on differential-geometric methods in physics*. Clausthal.

Hess, H. (1978b). An analogue of the Levi-Civita connection on symplectic manifolds. Preprint. Inst. of Theor. Phys., Freie Universität Berlin.

Hess, H. and Krauser, D. (1978). Lifting classes of principal bundles. Submitted to Rep. Math. Phys.

Hörmander, L. (1963). Fourier integral operators, I. *Acta Math*. Vol. 127, pp. 79-183.

Hurt, N.E. (1970). Remarks on Morse theory in canonical quantization. *J. Math. Phys*. Vol. 11, pp. 539-551.

Hurt, N.E. (1971a). A classification theory for quantizable dynamical systems. *Rep. Math. Phys*. Vol. 2, pp. 211-220.

Hurt, N.E. (1971b). Differential geometry of canonical quantization. *Ann. Inst. Henri Poincaré*. Vol. 14, pp. 153-170.

Hurt, N.E. (1972a). Topology of quantizable dynamical systems and the algebra of observables. *Ann. Inst. Henri Poincaré*. Vol. 16, pp. 203-217.

Hurt, N.E. (1972b). Homogeneous fibered and quantizable dynamical systems. *Ann. Inst. Henri Poincaré*. Vol. 16, pp. 219-222.

Joseph, A. (1970). Derivations of Lie brackets and canonical quantization. *Comm. Math. Phys.* Vol. 17, pp. 210-232.

Joseph, A. (1976). The algebraic method in representation theory (Enveloping Algebras). *Lecture Notes in Physics.* Vol. 50, pp. 95-109. Springer, New York.

Kazhdan, D., Kostant, B. and Sternberg, S. (1978). Hamiltonian group actions and dynamical systems of Calogero type. *Comm. Pure and Appl. Math.* Vol. 31, pp. 481-508.

Keller, J.B. (1958). Corrected Bohr-Sommerfeld quantum conditions for nonseparable systems. *Annals of Physics.* Vol. 4, pp. 180-188.

Kijowski, J. (1977). Geometric structure of quantization. *Lecture Notes in Mathematics.* Vol. 570, pp. 97-108. Springer, New York.

Kobayashi, S. and Nomizu, K. (1963). *Foundations of differential geometry.* Vol. 1. Interscience, New York.

Kolerov, G.I. (1978). Quantization and global properties of manifolds. *Ann. Inst. Henri Poincaré.* Sec. A. Vol. 27, pp. 137-154.

Kostant, B. (1970a). Quantization and unitary representations. *Lecture Notes in Mathematics.* Vol. 170, pp. 87-208, Springer, New York.

Kostant, B. (1970b). Orbits and quantization theory. *Actes Cong. Intern. Mat.* Nice. Vol. 2, pp. 395-400.

Kostant, B. (1970c). On certain unitary representations which arise from a quantization theory. *Lecture Notes in Physics.* Vol. 6, pp. 237-253. Springer, New York.

Kostant, B. (1972). Line bundles and prequantized Schrödinger equation. *Proceedings of the C.N.R.S. Colloquium on Group Theoretical Methods in Physics*, pp. IV1-IV22. Marseille.

Kostant, B. (1974). Symplectic spinors. *Symposia mathematica.* Vol. 14, pp. 139-152. Academic Press, London.

Kostant, B. (1975). On the definition of quantization, in *Géométrie symplectique et physique mathématique.* Coll. Int'l. C.N.R.S. No. 237, pp. 187-210. C.N.R.S. Paris.

Kostant, B. (1977). Graded manifolds, graded Lie theory, and prequantization. *Lecture Notes in Mathematics.* Vol. 570, pp. 177-306. Springer, New York.

Kostant, B. (1978). On Whittaker vectors and representation theory. Preprint. Dept. of Math. M.I.T.

Leray, J. (1975). Solutions asymptotiques et physique mathematique, in *Géométrie symplectique et physique mathématique*. Coll. Int'l. C.N.R.S. No. 237, pp. 253-276. C.N.R.S. Paris.

Lichnerowicz, A. (1943). Sur un principe de la dynamique lié à l'invariant intégral. *C.R. Acad. Sci.* Vol. 217, pp. 660-662.

Mackey, G.W. (1968). *Induced representations of groups and quantum mechanics*. Benjamin-Cummings. Reading, Mass.

Maslov, V.P. (1965). *Theory of perturbations and asymptotic methods* (in Russian). Moscow State University, Moscow.

Miller, J.G. (1976). Charge quantization and canonical quantization. *J. Math. Phys.* Vol. 17, pp. 643-646.

Nirenberg, L. (1957). A complex Frobenius Theorem. *Seminars on Analytic Functions*. Institute for Advanced Studies, Princeton, Vol. I, pp. I172-I189.

Onofri, E. (1974). Quantization theory for homogeneous Kähler manifolds. Preprint. Università di Parma.

Onofri, E. (1975). SO(n,2)-singular orbits and their quantization, in *Géométrie symplectique et physique mathématique*. Coll. Int'l. C.N.R.S. No. 237, pp. 155-161. C.N.R.S. Paris.

Onofri, E. (1976). Dynamical quantization of the Kepler manifold. *J. Math. Phys.* Vol. 17, pp. 401-408.

Onofri, E. (1977). V. Fock, 40 years later. *Lecture Notes in Mathematics*. Vol. 570, pp. 72-75. Springer, New York.

Onofri, E. and Pauri, M. (1972). Dynamical quantization. *J. Math. Phys.* Vol. 13, pp. 533-543.

Rawnsley, J.H. (1972). *Some applications of quantization.* Thesis, Oxford.

Rawnsley, J.H. (1974). De Sitter symplectic spaces and their quantization. *Proc. Camb. Phil. Soc.* Vol. 76, pp. 463-480.

Rawnsley, J.H. (1975). Diagonal quantization of Hamiltonians with periodic flows. Preprint, Mathematisches Institut, Bonn University.

Rawnsley, J.H. (1976). Coherent states and Kähler manifolds. Preprint. School of Theoretical Physics, Dublin Institute for Advanced Studies.

Rawnsley, J.H. (1977a). On the cohomology groups of a polarization and diagonal quantization. *Trans. Amer. Math. Soc.* Vol. 230, pp. 235-255.

Rawnsley, J.H. (1977b). A non-unitary pairing of two polarizations on the Kepler manifold. Preprint. School of Theoretical Physics, Dublin Institute for Advanced Studies.

Rawnsley, J.H. (1978a). Some properties of half-forms. *Lecture Notes in Mathematics*. Vol. 676, pp. 311-314. Springer, New York.

Rawnsley, J.H. (1978b). On the pairing of polarizations. *Comm. Math. Phys.* Vol. 58, pp. 1-8.

Rawnsley, J.H. (1978c). A non-unitary B-K-S pairing of polarizations. Preprint. School of Theoretical Physics. Dublin Institute for Advanced Studies.

Renuard, P. (1969). *Variétés symplectiques et quantification.* Thèse, Orsay.

Rosensteel, G. and Ihrig, E. (1978). Geometric quantization of the CM(3) model. To appear in *Ann. of Phys.*

Rothschild, L.P. and Wolf, J.A. (1974). Representations of semisimple groups associated to nilpotent orbits. *Ann. Sci. École Norm. Sup.* Vol. 1, pp. 155-174.

Schouten, J.A. (1954). *Ricci Calculus.* Springer, Berlin.

Segal, I.E. (1960). Quantization of non-linear systems. *J. Math. Phys.* Vol. 1, pp. 468-488.

Segal, I.E. (1974). Symplectic structures and the quantization problem for wave equations. *Symposia Mathematica.* Vol. 14, pp. 9-117.

Shale, D. and Stinespring, W.F. (1967). The quantum harmonic oscillator with hyperbolic phase space. *J. Funct. Anal.* Vol.1, pp. 492-502.

Simms, D.J. (1968). Lie groups and quantum mechanics. *Lecture Notes in Mathematics*. Vol. 52. Springer, Berlin.

Simms, D.J. (1972). Equivalence of Bohr-Sommerfeld, Kostant-Souriau and Pauli quantisation of the Kepler problem. *Proceedings of the C.N.R.S. Colloquium on Group Theoretical Methods in Physics*. Marseille. pp. IV23-IV28.

Simms, D.J. (1973a). Bohr-Sommerfeld orbits and quantizable symplectic manifolds. *Proc. Camb. Phil. Soc.* Vol. 73, pp. 489-491.

Simms, D.J. (1973b). Geometric quantisation of the harmonic oscillator with diagonalized Hamiltonian. *Proceedings of the Second International Colloquium on Group Theoretical Methods in Physics*. Nijmegen. Vol. 1, pp. 168-181.

Simms, D.J. (1974a). Geometric quantisation of energy levels in the Kepler problem. *Symposia Mathematica*. Vol. 14, pp. 125-137.

Simms, D.J. (1974b). The metalinear structure of the circular polarization in the plane. *Proceedings of the Third International Colloquium on Group Theoretical Methods in Physics*. C.N.R.S., Marseille, Vol. 1, pp. 72-76.

Simms, D.J. (1974c). Geometric quantisation of symplectic manifolds. *International Symposium on Mathematical Physics*. Warsaw.

Simms, D.J. (1975a). An application of polarization and half-forms. *Lecture Notes in Mathematics*. Vol. 466, pp. 195-197. Springer, New York.

Simms, D.J. (1975b). Metalinear structures and a geometric quantisation of the harmonic oscillator, in *Géométrie symplectique et physique mathématique*. Coll. Int'l. C.N.R.S. No. 237, pp. 163-174. C.N.R.S. Paris.

Simms, D.J. (1976). A survey of the applications of geometric quantisation. *Lecture Notes in Physics*. Vol. 50, pp. 111-116. Springer, New York.

Simms, D.J. (1977a). Serre duality for polarised symplectic manifolds. *Rep. Math. Phys*. Vol. 12, pp. 213-217.

Simms, D.J. (1977b). An outline of geometric quantisation (d'aprés Kostant). *Lecture Notes in Mathematics*. Vol. 570, pp. 1-10. Springer, New York.

Simms, D.J. (1978a). On the Schrödinger equation given by geometric quantisation. *Lecture Notes in Mathematics*. Vol. 676, pp. 351-355.

Simms, D.J. (1978b). Geometric quantisation and the Feynman integra. *C.N.R.S. Conference on Feynman integrals*. Marseille.

Simms, D.J. and Woodhouse, N.M.J. (1976). Lectures on geometric quantization. *Lecture Notes in Physics*. Vol. 53. Springer, New York.

Sławianowski, J.J. (1971). Quantum relations remaining valid on the classical level. *Reports Math. Phys*. Vol. 2, pp. 11-34.

Sławianowski, J.J. (1972). Geometry of van Vleck ensembles. *Reports Math. Phys*. Vol. 3, pp. 157-172.

Sławianowski, J.J. (1975). *Geometria przestrzeni fazowych.* P.W.N., Warsaw.

Śniatycki, J. (1974a). Prequantization of charge. *J. Math. Phys.* Vol. 15 (1974), pp. 619-620.

Śniatycki, J. (1974b). Bohr-Sommerfeld quantum systems. *Proceedings of the Third International Colloquium on Group Theoretical Methods in Physics.* C.N.R.S., Marseille, pp. 42-51.

Śniatycki, J. (1975a). Bohr-Sommerfeld conditions in geometric quantization. *Reports Math. Phys.* Vol. 7, pp. 303-311.

Śniatycki, J. (1975b). Wave functions relative to a real polarization. *Int. J. Theor. Phys.* Vol. 14, pp. 277-288.

Śniatycki, J. (1976). On wave functions in geometric quantization. *Lecture Notes in Physics.* Vol. 50, pp. 194-197. Springer, New York.

Śniatycki, J. (1977). On cohomology groups appearing in geometric quantization. *Lecture Notes in Mathematics.* Vol. 570, pp. 46-66. Springer, New York.

Śniatycki, J. (1978a). On geometric quantization of classical systems. *Mathematical Foundations of Quantum Theory.* Edited by A.R. Marlow, pp. 287-297. Academic Press, New York.

Śniatycki, J. (1978b). Application of geometric quantization in quantum mechanics. *Lecture Notes in Mathematics.* Vol. 676, pp. 357-367.

Śniatycki, J. (1978c). Further applications of geometric quantization. *Conference on differential-geometric methods in physics.* Clausthal.

Śniatycki, J. and Toporowski, S. (1977). On representation spaces in geometric quantization. *Int. J. Theor. Phys.* Vol. 16, pp. 615-633.

Śniatycki, J. and Tulczyjew, W.M. (1972). Canonical formulation of Newtonian dynamics. *Ann. Int. Henri Poincaré.* Vol. 16, pp. 23-27.

Souriau, J.-M. (1964). *Géométrie et relativité.* Hermann, Paris.

Souriau, J.-M. (1966). Quantification géométrique. *Commun. Math. Phys.* Vol. 1, pp. 374-398.

Souriau, J.-M. (1970). *Structure des systèmes dynamiques.* Dunod, Paris.

Souriau, J.-M. (1974a). Modèle de particule à spin dans le champ électromagnétique et gravitationnel. *Ann. Inst. Henri Poincaré.* Vol. 20, pp. 315-364.

Souriau, J.-M. (1974b). Sur le variété de Képler. *Symposia mathematica*. Vol. 14, pp. 343-360. Acedemic Press, London.

Souriau, J.-M. (1975). Mécanique statistique, groupes de Lie et cosmologie. *Géométrie symplectique et physique mathématique*. Coll. Int'l. C.N.R.S. No. 237, C.N.R.S. Paris.

Souriau, J.-M. (1976). Construction explicite de l'indice de Maslov, Applications. *Lecture Notes on Physics*. Vol. 50, pp. 117-148. Springer, New York.

Souriau, J.-M. (1977). Interprétation géométrique des états quantiques. *Lecture Notes in Mathematics*. Vol. 570, pp. 76-96. Springer, New York.

Sternberg, S. (1978a). On the role of field theories in our physical conception of geometry. *Lecture Notes in Mathematics*. Vol. 676, pp. 1-80. Springer, New York.

Sternberg, S. (1978b). Some recent results on the metaplectic representation. *Lecture Notes in Physics*. Vol. 79, pp. 117-143.

Sternberg, S. and Ungar, T. (1978). Classical and prequantized dynamics without Lagrangians or Hamiltonians. To appear in *Hadronic Journal*.

Sternberg, S. and Wolf, J.A. (1978). Hermitian Lie algebras and metaplectic representations. *Trans. Amer. Math. Soc.* Vol. 238, pp. 1-43.

Streater, R.F. (1966). Canonical quantization. *Comm. Math. Phys.* Vol. 2, pp. 354-374.

Streater, R.F. (1967). The representations of the oscillator group. *Comm. Math. Phys.* Vol. 4, pp. 217-236.

Tarski, J. (1976). Phase spaces for spin and their applications. *Acta Physica Austriaca*. Vol. 45, pp. 337-344.

Toporowski, S. (1976). *Representation spaces in geometric quantization for complete strongly admissible real polarizations*. M.Sc. Thesis, University of Calgary.

Tripathy, K.C. (1976). Dynamical prequantization, spectrum generating algebras and the classical Kepler and harmonic oscillator problems. *Lecture Notes in Physics*. Vol. 50, pp. 199-209. Springer, New York.

Van Hove, L. (1951). Sur le problème des relations entre les transformations unitaires de la mécanique quantique et les transformations canoniques de la mécanique classique. *Acad. Roy. Belgique, Bull. Cl. Sci.* Vol. 37, pp. 610-620.

Voros, A. (1975). The W.K.B.-Maslov method for nonseparable systems, in *Géométrie symplectique et physique mathématiques*. Coll. Int'l. C.N.R.S. No. 237, pp. 277-287, C.N.R.S. Paris.

Voros, A. (1977). Asymptotic $\hbar$-expansions of stationary quantum states. *Ann. Inst. Henri Poincaré.* Sec. A, Vol. 26, pp. 343-404.

Wallach, N.R. (1977). *Symplectic geometry and Fourier analysis,* in *Lie Groups:* History, Frontiers and Applications. Vol. 5, Math Sci Press, Brookline, Massachusetts.

Weinstein, A. (1971). Symplectic manifolds and their Lagrangian submanifolds. *Advances of Math.* Vol. 6, pp. 329-346.

Weinstein, A. (1974). Quasi-classical mechanics on spheres. *Symposia Mathematica.* Vol. 14, pp. 25-32. Academic Press, London.

Weinstein, A. (1975). Fourier integral operators, quantization, and the spectra of Riemannian manifolds. *Géométrie symplectique et physique mathématique.* Coll. Int'l. C.N.R.S. No. 237, pp. 289-298. C.N.R.S. Paris.

Weinstein, A. (1977). Lectures on symplectic manifolds. *Regional Conference Series in Mathematics;* No. 29. A.M.S., Providence, R.I.

Werth, J.E. (1975). Pre-quantisation for an arbitrary abelian structure group. *Int. J. Theor. Phys.* Vol. 1, pp. 183-190.

Werth, J.E. (1976). On quantizing A-bundles over Hamilton G-spaces. *Ann. Inst. Henri Poincaré.* Sec. A. Vol. 25, pp. 165-176.

Whittaker, E.T. (1959). *A treatise on the analytical dynamics of particles and rigid bodies.* Fourth Edition. Cambridge University Press, Cambridge.

Wolf, J.A. (1976a). Unitary representations of maximal parabolic subgroups of the classical groups. *Amer. Math. Soc. Memoir.* Vol. 180.

Wolf, J.A. (1976b). Conformal group, quantization, and the Kepler problem. *Lecture Notes in Physics.* Vol. 50, pp. 217-222. Springer, New York.

Wolf, J.A. (1978). Representations associated to minimal coadjoint orbits. *Lecture Notes in Mathematics.* Vol. 676, pp. 327-349, Springer, New York.

Woodhouse, N. (1976). Twistor theory and geometric quantization. *Lecture Notes in Physics.* Vol. 50, pp. 149-163. Springer, New York.

Woodhouse, N. (1977). Geometric quantization and quantum field theory in curved space-time. *Rep. Math. Phys.* Vol. 12, pp. 45-56.

Woodhouse, N. (1978). Geometric quantization and the W.K.B. approximation. *Lecture Notes in Mathematics.* Vol. 676, pp. 295-309. Springer, New York.

INDEX

GLOSSARY OF NOTATION

Symbol	Meaning	Pages
α	connection form	53, 181
$\mathscr{B}F$	linear frame bundle of F	64
$\tilde{\mathscr{B}}F$	metalinear frame bundle of F	64-65
$\mathscr{B}_{\omega}\mathscr{T}X$	symplectic frame bundle of X	87
$\tilde{\mathscr{B}}_{\omega}\mathscr{T}X$	metaplectic frame bundle of X	87-88
$\mathbf{C}$	complex numbers	
$\mathbf{C}^{\times}$	multiplicative group of non-zero complex numbers	
$C^{\infty}(X)$	smooth real-valued functions on X	
$\Gamma_{jk}^{\;\;i}$	Christoffel symbols	44
$D = F \cap \overline{F} \cap \mathscr{T}X$	involutive distribution	9
$D^{\mathbf{C}} = F \cap \overline{F}$	complexification of D	61
Div	covariant divergence	
d	exterior differential	
∇	covariant differential	54
	partial covariant differentiation	66
Δ	Laplace-Beltrami operator	133
$E = (F+\overline{F}) \cap \mathscr{T}X$	involutive distribution	9, 61
$E^{\mathbf{C}} = F + \overline{F}$	complexification of E	61
$\mathscr{E}$	quantum evolution space	28-29, 162
e	electric charge	
ε_{ijk}	permutation symbol	
$\underline{\xi}$	linear frame field	65
$\tilde{\underline{\xi}}$	metalinear frame field	65
ξ_f	Hamiltonian vector field of f	5, 39
F	polarization	8-9, 61
$\overline{F}$	complex conjugate of F	
f	electromagnetic field strength	

$Gl(n,\mathbb{C})$	complex general linear group	
g	metric tensor	
$\mathscr{H}$	representation space	10
$H^k(X,G)$	k'th Čech cohomology group with coefficients in the group G	
h	Planck's constant	
$\hbar$	$h/2\pi$	
hor ζ	horizontal part of ζ	53
Θ_Y	canonical one-form on $\mathscr{T}^*Y$	19, 40
$\mathscr{K}$	Blattner-Kostant-Sternberg kernel	80-81, 83
L	complex line bundle prequantization line bundle	52 55
$L^\times$	principal $C^\times$ bundle associated to L	52
$\mathscr{L}$	Lie derivative	
λ	section of L	
$\lambda^\#$	complex-valued function on $L^\times$ associated to λ	52
Λ	integral manifold of D	
$ML(n,\mathbb{C})$	complex metalinear group	64
$Mp(n,\mathbb{R})$	metaplectic group	87
ν	section of $\sqrt{\wedge^n F}$	
$\nu^\#$	complex-valued function on $\tilde{\mathscr{B}}F$ associated to ν	65
$\mathscr{P}$	prequantization map	6, 7, 51, 58-59
$\mathscr{P}_\omega\mathscr{T}X$	bundle of positive Lagrangian frames of X	88
$\tilde{\mathscr{P}}_\omega\mathscr{T}X$	bundle of metalinear positive Lagrangian frames of X	92
p_ζ	canonical momentum associated to the vector field ζ	20, 121
π	bundle projection	
$\mathscr{Q}$	quantization map	1, 14-15, 103-104, 108, 111

R	scalar curvature	133
$\mathbb{R}$	real numbers	
S	Bohr-Sommerfeld variety	71
$Sp(n,\mathbb{R})$	symplectic group	88
σ	element of $\mathscr{A}$	
$\underline{T}^t$	transpose of $\underline{T}$	
$\underline{T}^+$	Hermitian conjugate of $\underline{T}$	
T^k	k-torus	
$\mathscr{T}\phi_f{}^t$	derived map of $\phi_f{}^t$	
$\mathscr{T}X$	tangent bundle of X	
$\mathscr{T}^{\mathbb{C}}X$	complexified tangent bundle of X	
$\mathscr{T}^*X$	cotangent bundle of X	
tr	trace	
$\mathscr{U}$	intertwining operator	1-2, 3, 12, 77
ver ζ	vertical part of ζ	53
$\phi_f{}^t$	local one-parameter group of local canonical transformations generated by f	
X	phase space	5, 38
X/D	space of integral manifolds of D	
X/E	space of integral manifolds of E	
χ	holomorphic square root of determinant	65
Y	configuration space configuration space-time	19 27, 46
Z	evolution space Kaluza configuration space	28, 45 32
$\mathbb{Z}$	integers	
$\mathbb{Z}_2$	integers modulo 2	
ω	symplectic form; Lagrange bracket	38
ω^n	$\omega \wedge \cdots \wedge \omega$ (n times)	

ω_e	charged symplectic structure	<u>30</u>
$\sqrt{\wedge^n F}$	bundle associated to $\tilde{\mathscr{B}}F$ via χ	9-10, 11, 12, 64, <u>65</u>, 66, 101
$\langle\cdot,\cdot\rangle$	Hermitian inner product on L	<u>36</u>
$(\cdot\|\cdot)$	scalar product on $\mathscr{A}$	<u>69</u>
$[\cdot,\cdot]$	Poisson bracket; commutator	
$\wedge$	exterior product	
$\lrcorner$	left interior product	
$\otimes$	tensor product	
$\oplus$	direct sum	
$\|$	restriction	
$(\underline{u};\underline{v})$	symplectic frame	<u>87</u>